INSTITUTION OF CIVIL ENGINEERS

Bridge modification

Proceedings of the conference *Bridge modification* organized by the Institution of Civil Engineers and held in London on 23–24 March 1994

Edited by B. Pritchard

Thomas Telford

Organizing Committee: B. Pritchard (Chairman), W. S. Atkins Consultants Ltd and Colebrand Ltd, and D. Boyes and P. Dawe, Department of Transport.

Conference organized by the Institution of Civil Engineers and co-sponsored by the Department of Transport, the Institution of Highways and Transportation and the Institution of Structural Engineers.

Published on behalf of the organizers by Thomas Telford Services Ltd, Thomas Telford House, 1 Heron Quay, London E14 4JD

First published 1995

Distributors for Thomas Telford are
USA: American Society of Civil Engineers, Publications Sales Department, 345 East 47th Street, New York, NY 10017-2398
Japan: Maruzen Co. Ltd, Book Department, 3–10 Nihonbashi 2–chome, Chuo-ku, Tokyo 103
Australia: DA Books & Journals, 648 Whitehorse Road, Mitcham 3132, Victoria

A CIP catalogue record for this publication is available from the British Library.

Classification
Availability: Unrestricted
Content: Collected papers
Status: Authors' opinion
User: Civil and structural engineers

ISBN 0 7277 2028 7

© The Authors and the Institution of Civil Engineers, 1995, unless otherwise stated.

All rights, including translation, reserved. Except for fair copying, no part of this publication may be reproduced, stored in a retrieval system or transmitted in any form or by any means electronic, mechanical, photocopying, recording or otherwise, without the prior written permission of the Publications Manager, Publications Division, Thomas Telford Ltd, Thomas Telford House, 1 Heron Quay, London E14 4JD.

Papers or other contributions and the statements made or the opinions expressed therein are published on the understanding that the author of the contribution is solely responsible for the opinions expressed in it and that its publication does not necessarily imply that such statements and/or opinions are or reflect the views or opinions of the organizers or publishers.

Printed and bound in Great Britain by Redwood Books, Trowbridge, Wiltshire.

Contents

PART 1: WIDER AND LONGER

Policy

Policy traffic management and available options. D. S. BOYES	1
Why steel? G. W. OWENS and D. C. ILES	12

Design

The concrete options. G. SOMERVILLE	25
New ideas and innovations. A. C. G. HAYWARD	46
Widening bridges for the M1 and M6 motorways. A. M. PAUL and R. S. COOKE	56

Construction

Demolition and removal of existing prestressed concrete structures. P. LINDSELL	69
New and widened bridges for the M5 motorway. I. MARSHALL	77
The Kent approaches to the Dartford crossing. M. A. CHAMPION	90

Widening small and large

Widening of small bridges. A. E. NORFOLK and G. D. PETTITT	104
Rodenkirchen suspension bridge reconstruction and widening. R. HORNBY	125

PART 2: STRONGER AND SAFER

Policy

15 year programme and requirements. P. H. DAWE	139
The management of the assessment and strengthening programme. A. LEADBEATER	152

Case studies

Avonmouth bridge: assessment and strengthening. J. GILL, J. JAYASUNDARA and C. COCKSEDGE	162
Strengthening masonry arches. S. F. BROOMHEAD and G. W. CLARK	174
Bridge strengthening with minimum traffic disruption. B. PRITCHARD	185

Strengthening techniques

Plate bonding: a user's guide note. A. M. HENWOOD and K. J. O'CONNELL	204
Enhancing influences of compressive membrane action in bridge decks. A. E. LONG, J. KIRKPATRICK and G. I. B. RANKIN	217
Strengthening and modification of bridge supports. M. A. IMAM	228
Crossbeam replacement. M. S. CHUBB and I. L. KENNEDY REID	241

Alternatives to strengthening

Use of higher containment vehicle restraints (safety fences and barriers). C. WILSON	255
Is your strengthening really necessary? D. W. CULLINGTON and C. BEALES	270

Discussion on parts 1 and 2   285

Policy traffic management and available options,

D. S. BOYES, Project Director, Department of Transport

SYNOPSIS.
There has been a substantial growth in traffic in the UK since the motorway network was first built. The resultant demand for greater capacity is being increasingly met by plans to widen along the existing routes where traffic congestion is worst. The effect this has on the bridges, both in the present programme of work and the provision for future growth in demand, is considered.

INTRODUCTION
1. The provision of an effective transport network between the major centres of population, industry, and commerce is an essential pre-requisite for economic growth and prosperity in the community as a whole. It can be no surprise therefore that the unprecedented economic growth which has been achieved in the developed world in recent decades, has been accompanied by an equally unprecedented increase in demand for all modes of transport including air, rail and road.

2. This demand has led to increasingly heavier vehicles and a greater volume of traffic on UK roads, both of which have had, and will continue to have a direct effect on the nation's bridges. This paper principally addresses the implications of increased traffic volume but, before examining the consequences for the bridges, it will be worthwhile setting out a little of the background. Traffic volume is mainly regulated by the capacity of the road along which it travels and is only intermittently restrained by the bridges which intersect the road. The development of the Department of Transport's policy for accommodating future traffic growth in bridges is therefore strongly influenced by programmes such as that for motorway widening. This policy may be equally suitable for translation to other road programmes.

THE GROWTH OF ROAD TRAFFIC AND CONGESTION
3. In England, road transport has become the dominant means of moving both goods and people, catering for nearly 90 per cent of all inland needs. Traffic on the roads has grown within the period since 1982 by over 40 per cent, with particularly high growth on motorways, where traffic has more than doubled, and on trunk roads where traffic has increased by 50 per cent (1). It has been notoriously difficult to forecast these increases and, despite a

PART 1: POLICY

continuing investment in roads, the result has been growing traffic congestion. Adopting 1988 as a base year, traffic is now forecast to increase between 83 per cent and 142 per cent by the year 2025. Moreover, though the pivotal position that road transport has in relation to economic growth might be debatable, there is as yet no major change on the horizon that seems likely to upset the dominance of this form of transport.

4. Traffic congestion will increase in parallel with traffic growth in England unless a sustained and long term effort is made to overcome the problem. Congestion is grossly inconvenient for all road users, encouraging traffic to divert to less suitable roads, increasing the risk of accidents and reducing safety for the public at large. It also carries an environmental penalty by causing inefficient burning of fuel and increasing the level of exhaust emissions. But most importantly, it will be a source of increased cost and delays to British industry, making it less competitive with other countries in world markets.

5. The prospect of a growing traffic congestion spreading throughout the key routes in the national road network was not one the UK government would of course, willingly accept. Accordingly, the whole road programme was re-examined in the late 1980's and an expanded road programme was announced (2) adding over 2700 miles of new or widened roads to the national trunk road network over the following decade. In the intervening period there have been additions, changes and modification to this programme and on the 5 August 1993, John MacGregor, Secretary of State for Transport announced a comprehensive review of the Road Programme. It is clear however that a substantial programme of road widening and construction will be with us for many years ahead.

THE CONSEQUENCES FOR BRIDGES
6. It is important when devising any national programme for the provision of road schemes that there should be an overall balance. Nevertheless, having established that traffic demands are out-stripping the existing capacity of the road network, a strong case can be made for increasing capacity along the broad routes followed by the existing motorways. Improvement along existing routes avoids the creation of new traffic corridors and provides an opportunity to upgrade the environmental measures incorporated into the works when they were built. Widening of these routes will also enable attention to be more sharply focused on where congestion is greatest and the needs of industry best served and experience has shown that widening schemes on existing roads can be completed in less time than the building of new roads where the time to completion can be substantially extended due to statutory procedures.

7. It is usually the case that where a substantial advantage exists a substantial disadvantage will also occur. In the case of motorway widening the disadvantage is the need to maintain the flow of a large traffic volume while the works are undertaken. Many motorways are already operating at well above their design year flows and 80-100,000 vehicles a day is not unusual with some routes as

high as 150,000 vehicles per day. Imposing restrictions at this level causes frustration, adding to the accident risks and inevitably giving rise to considerably increased costs for delays. Thus, there will be a strong impetus towards adopting traffic management measures which will impose the least possible restrictions on the existing traffic.

8. The decision to implement a road widening programme and to cause the least disruption to traffic means that careful planning and clear policies are paramount. Two principal issues can be identified in this context which will have a considerable influence on the available options for either modifying or replacing the bridges within the present programme. They are:

a) bridges are themselves a constraint on the widening methods to be employed.

b) the choice of replacement bridges will be geared to construction in a live traffic situation and not necessarily to minimum initial capital cost.

9. Bridges built in the original motorway programme of the 1960's and 70's have little if any flexibility to accommodate the present widening proposals which means many bridges less than 30 years old will need to be demolished and rebuilt. Bridges traditionally account for approximately 25 per cent of total scheme costs, to which must now be added therefore, the costs of any necessary demolition in the present widening process and the attendant costs of traffic delay. The result is that the format for bridge modification and replacement will clearly have a serious impact on the cost and pace of the present programme. There will be a need for an innovative approach to bridge design, especially overbridges, to minimise this impact and no doubt a variety of bridge formats will be adopted. Moreover, it would be prudent at this stage to re-examine the basis on which bridges are to be provided, as part of the present proposals, in order that they might have more flexibility for accommodating changes which may be required in the future. A sensible approach would be to provide bridges that will not require modification or cause disruption if further widening is needed, without increasing the present level of costs.

METHODS OF WIDENING, TRAFFIC AND STRUCTURAL OPTIONS

10. Numerous variations on the theme of motorway widening have been considered and developed in he search for a perfect system that would eliminate some, if not all, of the traffic management problems. These range from minimum or zero land-take solutions, where the hardshoulder is dropped at pinch points such as through the bridges, to completely off-line solutions such as link roads. All of these variations can be grouped into three basic methods, symmetrical, asymmetrical and parallel widening. Each has a different impact on traffic management, affecting both the ultimate choice and the options for the structures (3). These methods are shown diagrammatically in Fig. 1 for widening from dual-three lanes to dual-four lanes.

11. The simplest form of widening is the symmetrical method which will add road width equally to both

PART 1: POLICY

carriageways. In this method the land-take can be kept to a minimum, or even zero, with the existing hardshoulder reconstructed as a running lane. The construction period will obviously be reduced if widening can take place simultaneously on each side but, if it is necessary to keep three traffic lanes open at all times, then the widening will have to be undertaken in phases. Inevitably this will mean the loss of one of the original running lanes during the construction period as a safety zone and, with two lanes remaining, a third lane will have to run in contraflow on the unaffected carriageway with the hardshoulder brought into service to provide three lanes in the opposite direction. This can be quite costly in terms of traffic disruption, if the traffic is already congested, and may be so high that such a solution is not economically viable. The symmetrical method of widening is therefore more appropriate for schemes where either traffic congestion is not yet occurring or where other constraints, such as urban development, make it the only practicable choice.

12. A new hardshoulder can be built alongside the new alignment either within existing motorway boundaries or with some extra land acquisition. It is possible that some of the structures could be retained with this method of widening, provided dropping of the hardshoulder through the bridge is acceptable. At underbridges this would mean that traffic will have to travel close to the edges of the deck and an assessment will then be required to check that the edge of the deck is stiff enough to carry the imposed loading. Dropping of the hardshoulder through the overbridges will bring most side-piers into the impact zone and these may then be unable to meet the present impact criteria. Extra protection will be required and this would most likely be in the form of the recently introduced solid concrete wall.

13. The method of asymmetrical widening is most likely to be adopted where a particular constraint exists along one side of a motorway. In this method the old and new carriageways follow similar alignments along one edge, while the further edge will lie well beyond the original alignment. The motorway is effectively moved laterally by two lane widths. Constructing the new lanes and hardshoulder along the widened edge would be constrained by the existing bridges and it may therefore be necessary to demolish existing overbridges before the carriageway works begin. There would be a difficulty in replacing the bridges immediately because the ultimate position of the central reserve would be in part of the existing live carriageway, making the construction of an intermediate pier problematic. Alternatives range from a 40m plus single span, through temporary bridges or road closures, to reverting from asymmetrical to symmetrical widening through the bridge sites.

14. The sequence of traffic management is more complicated than in symmetrical widening because of the need to reposition the central reserve. When extending the outer edge of the carriageway, it is likely that only two running lanes can be kept open on the widening side. Therefore, similarly to symmetrical widening, in order to maintain the traffic capacity a third lane would have to run in

contraflow on the other carriageway whilst maintaining three lanes in the opposite direction by using the hardshoulder. At the stage of replacing the central reserve, a working space of about four running lanes will be required which will mean working to tight allowances if three running lanes are to be kept open at all times. Furthermore, the need to maintain traffic capacity during the different stages of the work will be complicated by being undertaken in association with work on adjacent sections as well as carriageway closures for the lifting-in of bridge beams and decks.

15. Parallel widening requires the acquisition of sufficient land to construct a completely new wider carriageway alongside the existing road without the need to overlie any part of it. Once the new carriageway is completed, traffic can be managed between the three available carriageways whilst the existing motorway is reconfigured to form a second wider carriageway. The advantage of this method of widening is that it will eliminate virtually all restrictions on traffic during the progress of the works. Though traffic management will need to be undertaken in a number of phases it will avoid the loss of any running lanes with the exception of limited closures for the purposes of bridge erection.

16. The technique is constrained by the existing bridges similarly to the asymmetrical method of widening. Another option is available for the overbridges however, assuming the side road can be re-aligned. An intermediate bridge pier can be constructed between the edge of the existing motorway and the new carriageway, in association with the construction of the latter. A deck can then be landed on this, and new abutments at the verges, before the old bridge is demolished to allow completion of the new carriageway.

17. If this new pier is placed at a minimum clearance from the edge of the existing motorway, the reconstructed carriageway will spread over all or part of both of the existing carriageways. Obviously, the pier can be placed at a greater clearance so as to avoid this, in which case traffic on one of the existing carriageways can remain in its normal configuration whilst traffic can be switched from the other to the new carriageway. The reconstruction process is then isolated from any conflict with traffic management procedures until completion of the works. Ideally, the bridge span over what was the original motorway should be a single span to avoid further traffic disruption, but it too will now be 40m plus.

STRATEGY FOR BRIDGES
18. This paper has described how road traffic has grown, and will continue to grow, and how this leads to increasing traffic congestion. The benefits of making extra road provision by virtue of motorway widening have been examined and the various methods of widening have been illustrated. A major issue is that many overbridges - and some underbridges - will have to be replaced when widening is carried out. Bridges are designed to last for 120 years and their replacement, typically only 30 years after they were first built, means that not only is motorway widening incurring heavy additional costs due to bridge demolition and the traffic disruption that is caused but also a

PART 1: POLICY

valuable national asset is being lost prematurely.

19. When the original motorway network was constructed there was no grand perception that widening would ever be needed and, in those circumstances, no perception that bridges would be replaced in so short a time. In hindsight that is seen to be short-sighted and it is now considered appropriate to ensure that bridges are provided with as much flexibility to meet future needs as possible. Cost estimates for alternative designs, including those for bridges, should take into account the capital costs of construction as well as the discounted costs of any demolition and the associated traffic disruption that may occur.

20. Projected traffic forecasts show that many motorways will reach congestion levels again within 30 years of being widened - the period over which the economics of all highway schemes are assessed to justify investment in them. In some locations, should the higher range of forecast traffic growth be realised, the design flow may be exceeded within the 15 year design period. Both these periods are so much shorter than the 120 year design life for bridges that, if we are to make any sense of the latter, there is a positive need to ensure that bridge replacement is not initiated by actions emanating from the former. Simply put, it would be eminently sensible to adopt a policy whereby premature demolition of bridges could be avoided if widening should become necessary in the future. And of course, such a policy should be demonstrably economically sound.

21. Although a number of the present overbridges will no longer be serviceable, most will be replaced because they cannot accommodate the wider carriageways which are now being demanded. Their piers or abutments are too close to the existing carriageway to allow a new running lane to be constructed alongside. But positioning the side piers in the replacement bridges so as to make an allowance for future widening, bearing in mind that main spans will already be greater than hitherto, would create a structurally less efficient span arrangement as well as an unpleasing appearance. The solution is to avoid side spans altogether. In the majority of cases this will mean the preferred option will be a two-span bridge with the deck supported by a centrally positioned pier and at the top of side-slopes at the ends.

22. There are a number of other features which need to be examined in pursuing this option, principally land acquisition, environment and aesthetics but most importantly cost. In evaluating the costs, the assessment would normally be taken over the 30 year period adopted for the assessment of the economics for highway schemes and would entail a comparison of the discounted costs for all the possible alternative designs. Furthermore, the discounted costs should include not only the basic capital costs of the bridge but also the full costs of demolition, alteration and any traffic delay costs that can be attributed to the bridges within the 30 years.

23. Fortunately, the two span open side slope option coincides with aesthetic advice on bridges which advocates avoiding bulky abutments or fussy side-piers which obstruct clear views of the land-form through the bridge. A clean simple line of the deck imposed on the landscape is to be preferred and as the following examples demonstrate this type of bridge can be contained within the normal widening land-take to give the desired flexibility.

24. This can be best illustrated by considering two specific examples. The first example is that of replacement overbridges in a symmetrical widening scheme where only one lane is being added to each carriageway (Fig 2). The choice of a replacement would lie between a bridge with piers or abutments close to the edge of the new carriageway and a bridge with the supports set further apart, close to the top of side slopes. It will be obvious from the diagram that, if it were necessary, the two-span bridge with open side-slopes would allow the carriageways to be widened at a later stage without the need for substantial bridge modification, whereas the other would not. There should be virtually no difference in the land-take for these bridges, as both could be constructed within the overall motorway width and there should also be little difference in cost. Spans will be approximately 35m.

25. The second example considers parallel widening and again the choice lies between two alternative bridges (Fig 3). In this case the new abutment adjacent to the existing carriageways is restrained in position by the need to maintain traffic on these carriageways while the bridge and the new carriageway is being built. The left-hand abutment adjacent to the new carriageway can be built either at minimum clearance or at the top of open side-slopes. The former design has a very asymmetric appearance because the span over the new carriageway is only about half the length of the other. By stretching the short span to the top of side-slopes and bringing the earthworks over part of the redundant carriageway in the right-hand span, the appearance is improved by imparting a greater symmetry. Both bridges can be built within the overall motorway width but it is obvious the open side-slope version provides greater flexibility for future widening if that were to prove necessary, without the need to re-build the bridge. The spans of the latter would vary between about 38m and 43m.

26. Underbridges which carry the motorway or trunk road present a different problem. The issues for the existing underbridges affected by the present widening programme have already been addressed. The major traffic disruption is likely to be caused to vehicles on the bridge - rather than those passing under it. The issue of making immediate provision of greater width of underbridge deck to accommodate these vehicles, in anticipation of some speculative future widening of schemes, may not be economically viable. Fortunately the bridges lend themselves to extending the decks at a later stage without excessive traffic disruption. In those few cases where one motorway passes over another, or other heavily trafficked road, it may be necessary to consider providing not only extra length in the span but also width across the deck. Both aspects would be to avoid traffic disruption on

PART 1: POLICY

the lower road caused by the work of reconstruction at some later stage. These would be special cases for the underbridges and need to be carefully assessed, though the criteria to be used should be similar to those for overbridges.

EPILOGUE

27. It seems certain that traffic congestion, both on motorways and other roads, will continue to be a problem for many years ahead. The demand for increased capacity on the roads is one that the UK government cannot afford to stand aside from whereas there are substantial benefits to be gained in matching it. Nonetheless, the provision of extra capacity is a costly exercise and the search for value-for-money will always take high priority in delivering an overall programme.

28. Beyond a certain period of time, it is not economically sensible to make an immediate provision for extra capacity. But, if growth in demand continues, then it may be necessary to re-widen some motorways at some time in the future. Experience has shown that the bridges have a major impact on the progress of such work, causing delay and adding to the cost of schemes. It would therefore be prudent, in building bridges with a 120 year design life, to incorporate sufficient flexibility into their design to accommodate some future provision and avoid expensive demolition and replacement of bridges. This is especially true if they can be contained within present land-take and built at no extra cost to the public purse.

Building bridges which have a pleasing appearance has always carried considerable weight with the Department of Transport, who have regularly consulted with the Royal Fine Arts Commission on a wide range of structures and have offered guidance and advice to designers (4). Since the advent of the motorway widening programme there has been a growing awareness of the impact of bridge appearance and how this equates to the provision being made in the present programme of motorway widening has been briefly mentioned. Ministers have now expressed their considerable interest in the subject and as a consequence it seems likely that a number of design competitions for major bridges, and some standard bridges, will be forthcoming. Taken together with the policy being advocated, this should ensure that not only will bridges match the longevity inherent in their design with accommodating traffic growth but also that they will enhance the environment of which they are a part.

References.

1. Transport statistics Great Britain 1993: HMSO.

2. Roads for Prosperity CM693 1989: HMSO.

3. The Good Roads Guide Motorway Widening, Widening Options and Techniques HA62/92: Department of Transport.

4. Bridge Appearance: Department of Transport 1964.

Widened

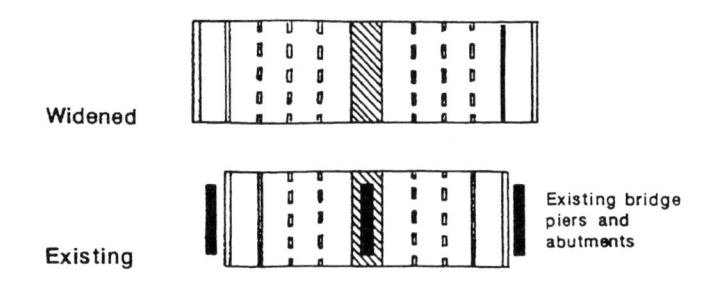

Existing — Existing bridge piers and abutments

Symmetrical

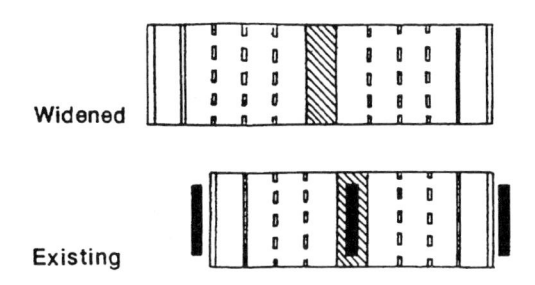

Widened

Existing

Asymmetrical

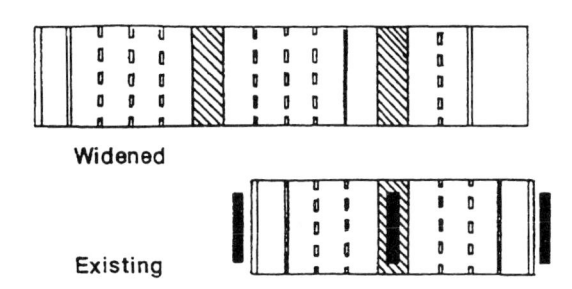

Widened

Existing

Parallel

Fig 1 Methods of On-line Widening

PART 1: POLICY

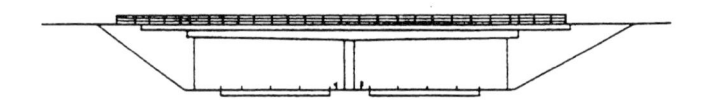

Existing bridge

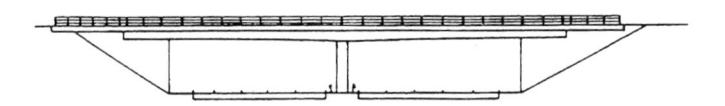

Abutment design – leaves no flexibility for further widening on either side

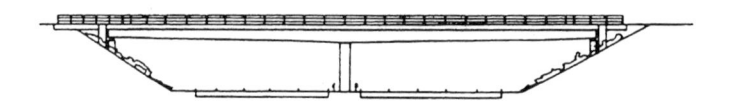

Open side slope design – gives flexibility for further widening on both sides

by removing side slopes

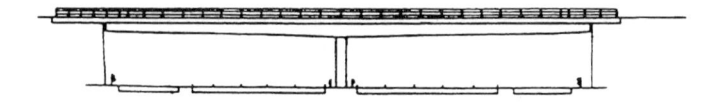

Fig. 2 Symmetrical Widening Dual 3 lanes to dual 4

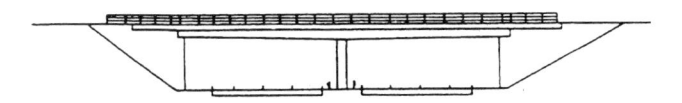

Existing bridge

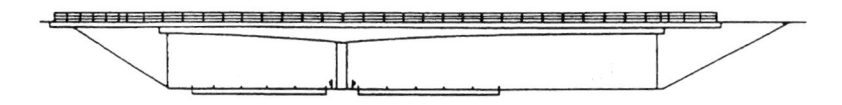

Abutment design – leaves no flexibility for
further widening on the left hand carriageway

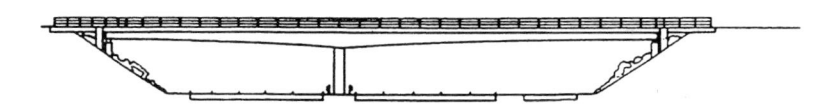

Open slope design – gives flexibility
for further widening

by removing side slopes

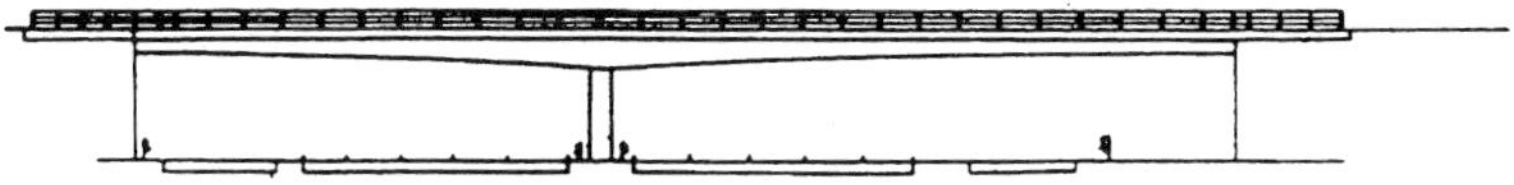

Fig 3 Parallel Widening Dual 3 lanes to dual 4

Why steel?

G. W. OWENS, Director, and D. C. ILES, Manager, The Steel Construction Institute

SYNOPSIS This paper summarises the principal advantages of specifying steel for bridge modifications. While primarily addressing the current state-of-the-art it also draws attention to likely future developments that could provide further benefit to the specifier.

INTRODUCTION
1. As the amount of traffic on our roads increases, so the bridges that carry that traffic are having to be modified to meet the changing demands. Existing bridges may be modified, in strength or size, or they may need to be replaced by new structures.
2. There are several reasons why steel is particularly suitable as the construction material for these modifications and replacements.

QUALITY
3. Fabricated steelwork for bridges is a quality product manufactured from a quality material.
4. For bridges an effective range of modern steels that includes weathering steel and the traditional high tensile steel (strength grades 355 N/mm^2) is available to the new European Standards[1,2]. During the 1970's the demands of the offshore sector in the North Sea improved the cleanliness, toughness and weldability of structural steels and these advances have been of considerable benefit to steel bridges.
5. Bridge steelwork is fabricated in a controlled environment using a stable, well-trained workforce and modern equipment. Most bridge fabricators in the UK are accredited under a recognised Quality Assurance scheme, usually one that has been specifically developed for the needs of the steel construction industry and its clients[3].
6. The prefabricated components are now fabricated to a high degree of accuracy and can be relied upon to fit on site - first time.

ECONOMY
7. Structural steel and fabricated steelwork have both improved their relative economy over the last two decades. The major improvements took place in the late 70's and early 80's but the relative advantages achieved then have been maintained subsequently.
8. Improvements in steel production, including the development of continuous casting and increased automation has improved yields and enabled massive reductions in manning levels. Figure 1 shows the improvement in productivity.

9. Improvements in fabrication (see Figure 2) have been less dramatic but still very worthwhile. Improvements in cutting, holing, handling and welding have all been important and have been matched by improvements in information flow and rationalisation of detailed design.

DESIGN

10. Considerable advances have also been achieved in both the conceptual and detailed design of steel and composite bridges.

11. The research effort which ensued after the traumas of the box girder failures of the 70's led to a considerable improvement in our understanding of plate behaviour. Those events are now well behind us and our understanding has matured, for the most part, into usable and economical design rules.

12. This stability of detailed design environment has enabled more attention to be devoted to conceptual design. It has in turn led to developments in long span construction[4,5] and, of more relevance to bridge modification, a range of potential solutions for replacement motorway overbridges[6,7].

13. This range of solutions covers seven types of steel or composite bridges (see Figure 3). A case study for each solution was carried out by a Consulting Engineer, and the SCI publication[6] expresses the contribution of new ideas for replacement overbridges from each designer.

14. The solutions are adaptable to different widening options, symmetric, asymmetric and parallel, and to the largest spans which provide flexibility for any future changes in highway layout[7]. Typical elevations are shown in Figure 4.

15. In addition to being able to consider a wide range of usable forms of cross-section, it is also possible to adopt different approaches to reducing the depth of construction, see Figure 5.

16. The most straightforward technique to reduce depth is to use continuity over intermediate supports. Continuity is easily achieved with steel construction and improves economy as well as offering the scope to minimise construction depth. Indeed, continuity is likely to be demanded in modern construction to minimise the numbers of roadway joints and their consequent maintenance requirements.

17. With plate girders it is now relatively cheap to vary the depth of the girders along its length. This is most economically achieved by using tapered lengths of girder, with straight flanges - effectively haunched girder construction. Alternatively, for a small additional premium, curved bottom flanges can be specified, which is often considered to improve appearance.

18. Finally, it is possible to use half-through girder construction. In such structures the effective depth of construction (road surface to soffit) becomes independent of longitudinal span and for most over-bridges will be less than a metre.

19. Steel construction can also readily accommodate high skews. With appropriate bracing arrangements a plate girder structure has low torsional stiffness and torsional stresses can therefore be minimised.

20. Thus recent developments in conceptual design of steel bridges offer a range of viable solutions. The very variety of these aesthetically pleasing structures adds to the richness of our built environment.

PART 1: POLICY

SPEED OF CONSTRUCTION
21. Any modification to existing bridges or highways must clearly be carefully planned to minimise disruption to traffic. For example, with the replacement of motorway overbridges, the primary objective must be to reduce to an absolute minimum the period of time that the motorway itself is partially or totally closed. A secondary consideration will be to reopen the overbridge as soon as possible.
22. Steel bridge construction is very well suited to fast on-site construction. All element fabrication is carried out remote from the site. It is frequently possible to pre-assemble the complete steel superstructure and use a high-capacity (500 tonne plus) crane to lift it into its final position during a limited possession. Figure 6 shows a typical example. Pre-assembly can normally be carried out either close to the final position or some distance away, using special transporters to move the completed structure to site.
23. In the motorway widening study[6], initial planning studies indicated that all the schemes could be lifted into place with a single possession of 6 to 12 hours, Figure 8 shows a typical programme.
24. Alternatively the elements are assembled at site into a small number of components, each weighing less than 100 tonnes. As such they can be lifted by mobile telescopic cranes of the type that can travel along the public highway yet be ready to lift within a very short time of arrival.

DURABILITY
25. The inspection and maintenance of steel bridges is now a well known and mature technology.
26. Conventional steels do require inspection and painting at regular intervals. A particular benefit of steel is that any deterioration or corrosion will always occur from the outside and will thus be visible as soon as it starts. Inspection, identification and remedial action for any problems are therefore straightforward.
27. As an alternative, the use of weathering steels in appropriate non-marine and non-polluted environments will lead to structures that require a minimum of maintenance.

ADAPTABILITY
28. Previous experience would suggest that there are likely to be further extrapolations in loading or other service requirements during the lifetime of our newer bridge stock. However, steel is an adaptable structural medium and its use builds in an inherent adaptability into the structure which might prove to be invaluable in the future.
29. With suitable ingenuity it is possible to devise and implement strengthening systems for increases in most of the load effects in the primary elements, secondary structure and connections, even for major suspension bridges[8]. Sometimes conventional connections of welded and bolted connections will suffice; sometimes special fasteners and adhesives may have to be used. Later papers in this conference highlight what can be achieved for bridges; reference 9 summarises opportunities in building structures.
30. The ultimate adaptability is to demolish the structure and rebuild to the new specification. A steel structure may readily and safely be cut up by conventional means. The resultant scrap has a positive net worth and may be recycled with a minimum expenditure of energy - a considerable environmental benefit.

TECHNICAL SUPPORT

31. A less tangible but nonetheless significant reason for specifying steel for bridge construction is the strength of technical support that is available from the structural steelwork community. Coordinating their efforts through the Steel Construction Industry Federation (SCIF), British Steel, the British Constructional Steelwork Association and The Steel Construction Institute collectively provide a comprehensive and high quality technical support for all users of steel.

FUTURE DEVELOPMENTS

32. The previous sections of this paper have highlighted the varied reasons for using steel for bridge structures. The strength of these arguments is demonstrated by the strong and growing market share of steel in bridge construction. That is shown in Figure 9. In recent motorway widening contracts over 90% of overbridges have been of composite construction.

33. It is possible to identify a range of future developments that should increase the benefits of using steel.

Integral Construction

34. There is a rapidly growing interest in the use of integral construction, whereby expansion joints and bearings are eradicated, thus substantially reducing future maintenance costs. Thermal actions are partly resisted within the structure and partly accommodated by movements of the embankments.

35. Initial studies have shown that thermal forces are significantly less in a steel superstructure[10].

36. Integral construction lends itself to the use of steel substructures, which can be made to achieve full continuity between sub- and super-structure. The use of tubular piles and installation techniques developed for the offshore sector could readily be applied to bridge piers and abutments, is shown in Figure 10.

Higher strength, more weldable steels

37. Higher strength, quenched and tempered steels have been developed and are used in bridge construction elsewhere in the world, notably in Japan. Discussions are currently in hand to investigate the means by which the Eurocodes may be expanded to encompass their use.

More automated fabrication

38. The manufacture of the basic plate girder is now a highly automated, cost-effective process. In time the same degree of automation will also be applied to completing the fabrication, including the robotic positioning and welding of stiffeners and other fitments.

Computer integrated design and information flows

39. Substantial use is already made of computers for design, detailing, drawing and management information. The next step will be a fully integrated system with savings in time, checking, and material waste.

Maintenance

40. The present procedures for the inspection, maintenance and painting of steel structures have changed little in recent years. Timescales, for example three weeks for maintenance painting, are no longer appropriate for today's environment. Further development of access

PART 1: POLICY

arrangements and protection systems for painting, methods of preparation, paints and application methods could reduce these timescales considerably.

CONCLUSIONS
  41. Steel has several significant advantages as a material for bridge modification and replacement.
- Fabricated steelwork is a quality product.
- Its relative economy has improved considerably in the last two decades.
- A variety of design concepts is available for both long and shorter span bridges.
- Fast construction on site can readily be achieved.
- Its maintenance is a well established technology.
- The structure may be modified in the future.
- It is well supported by its industry.

  42. Advances in the following areas will further enhance steel's attractiveness.
- The development of integral construction.
- Increases in automation of fabrication.
- Evolution of computer integrated manufacture.
- Developments in initial and maintenance painting.

REFERENCES
1. BS EN 10 025: 1990, Specification for hot rolled products of non-alloy structural steels and their technical delivery conditions.

2. BS EN 10 155: 1993, Structural steels with improved atmospheric corrosion resistance - technical delivery conditions.

3. Bridgework Schedules, Steel Construction Quality Assurance Scheme, London, (to be published in March 1994).

4. VIRLOGEUX, M. Normandie Bridge, Paper 10150, Proc. Inst. Civil Engineers, August 1993.

5. GIMSING, N.J., The Storebælt East Bridge, International Symposium - Bridges in Steel, ECCS, Paris, 1992.

6. ILES, D C (ed). *Replacement Steel Bridges for Motorway Widening*. P204, The Steel Construction Institute, 1992.

7. ILES, D C. *Motorway Widening: Steel Bridges for Wider Highway Layouts*. P208 The Steel Construction Institute 1992.

8. Strengthening and refurbishment of Severn Crossing, Papers 9845 to 9849, Proc. Inst. Civil Engineers, February 1992.

9. TREBILCOCK, P. *Adaptability in Steel*. P139, The Steel Construction Institute (to be published in 1994).

10. HAMBLY, E & OWENS, G W. Jointless Steel Viaducts. IABSE Henderson Colloquium, Cambridge, July 1993.

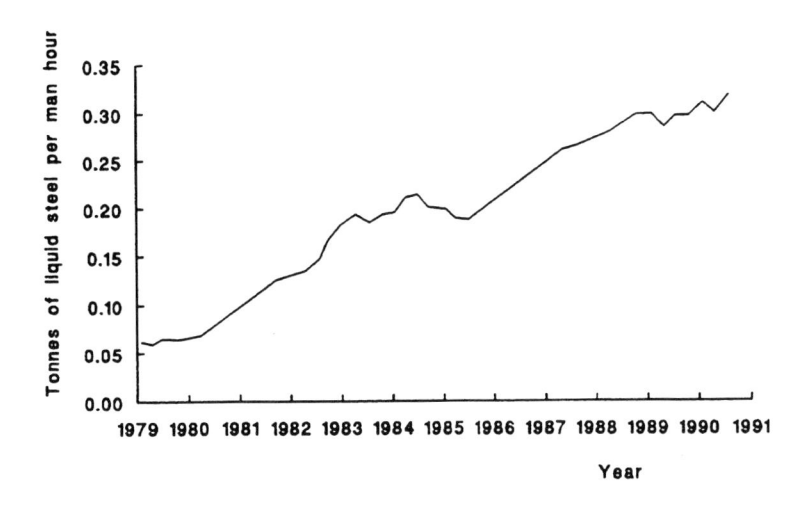

Figure 1 Improvements in relative productivity of steel manufacture

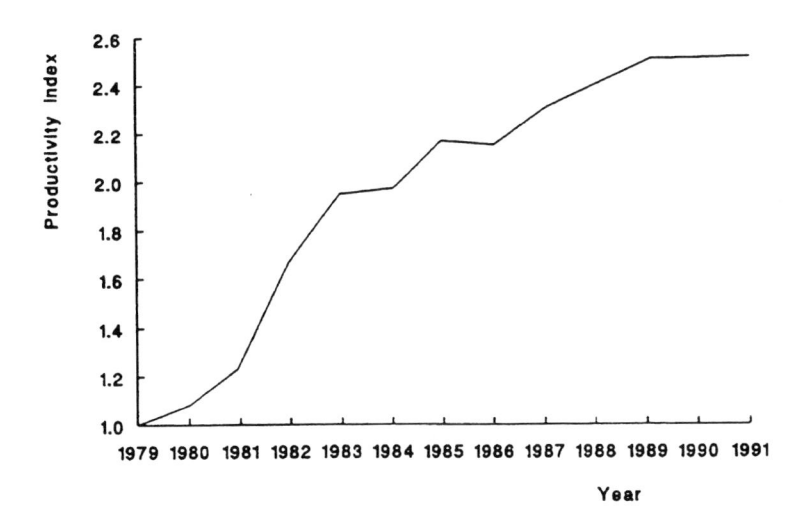

Figure 2 Improvements in relative productivity of steel fabrication

PART 1: POLICY

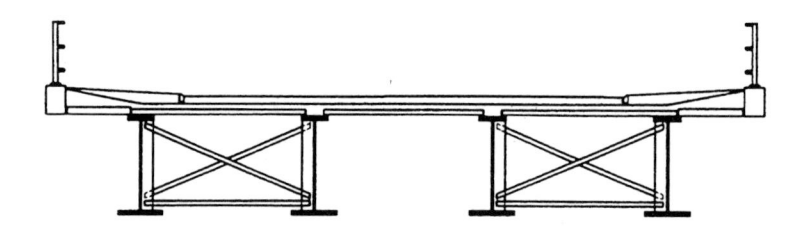

(a) Composite bridge with permanent steel shutter

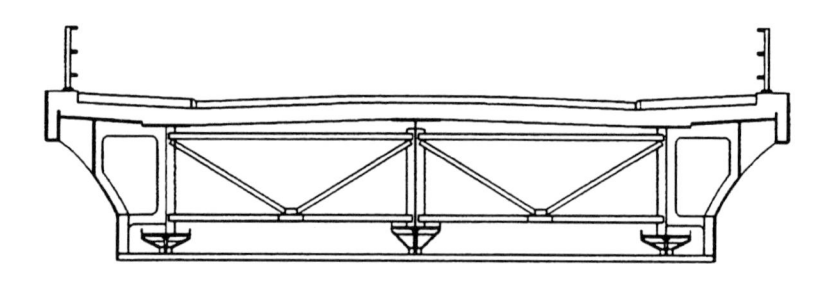

(b) Fully enclosed composite plate girder bridge

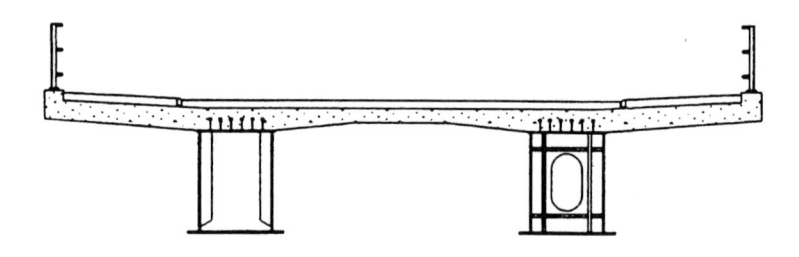

(c) Composite bridge using precast deck panels

Figure 3(i)   Cross-sections of replacement bridges for motorway widening

(d) Half-through plate girder bridge

(e) Half-through truss bridge

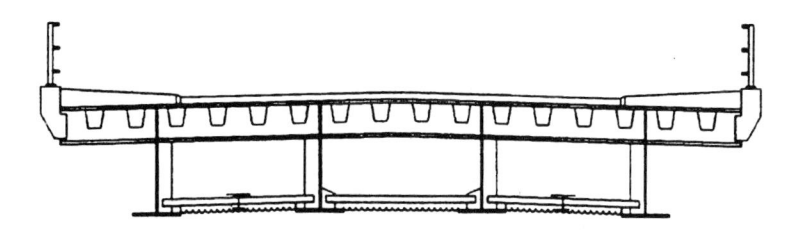

(f) Plate girder bridge with orthotropic deck

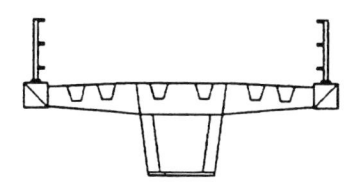

(g) Farm access bridge using steel box and orthotropic deck

Figure 3(ii)   Cross-sections of replacement bridges for motorway widening

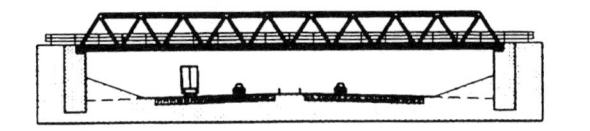

PART 1: POLICY

	Spans (m)
(a) Single span - solid abutments	45 to 60
(b) Three spans - open side spans	 15 to 20 45 to 60 15 to 20
(c) Two spans - usually following parallel widening	35 to 45 35 to 45

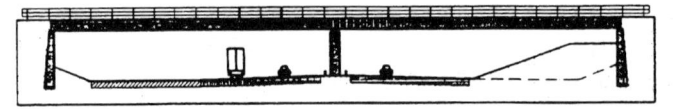

Figure 4   Span ranges for different types of motorway widening

(a) Simple spans

(b) Continuity

(c) Variable depth

(d) Half through

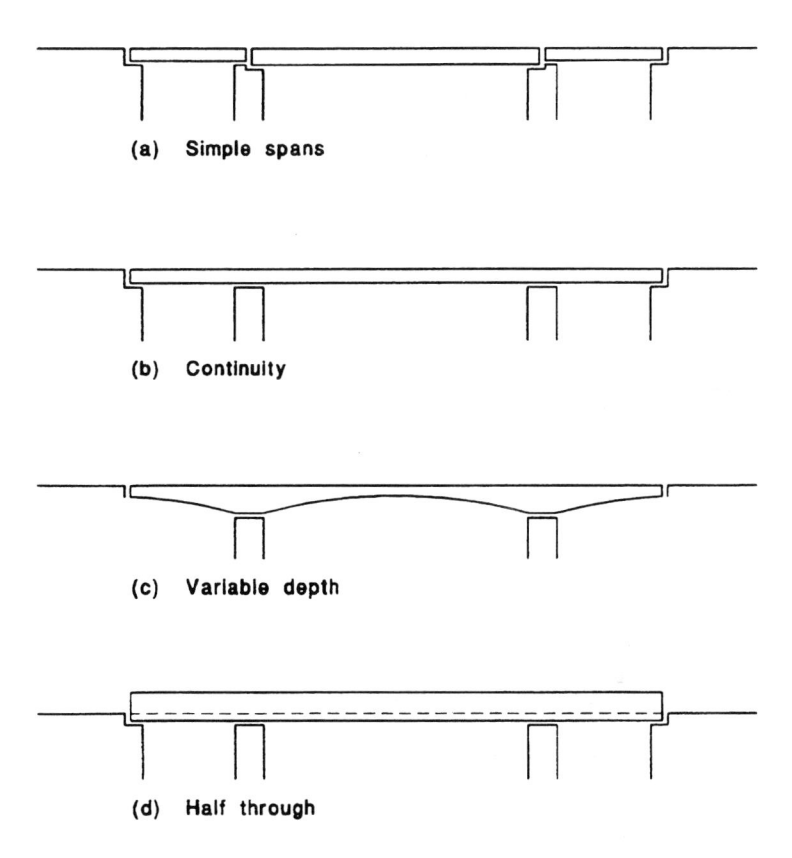

Figure 5 Methods of reducing structural depth

PART 1: POLICY

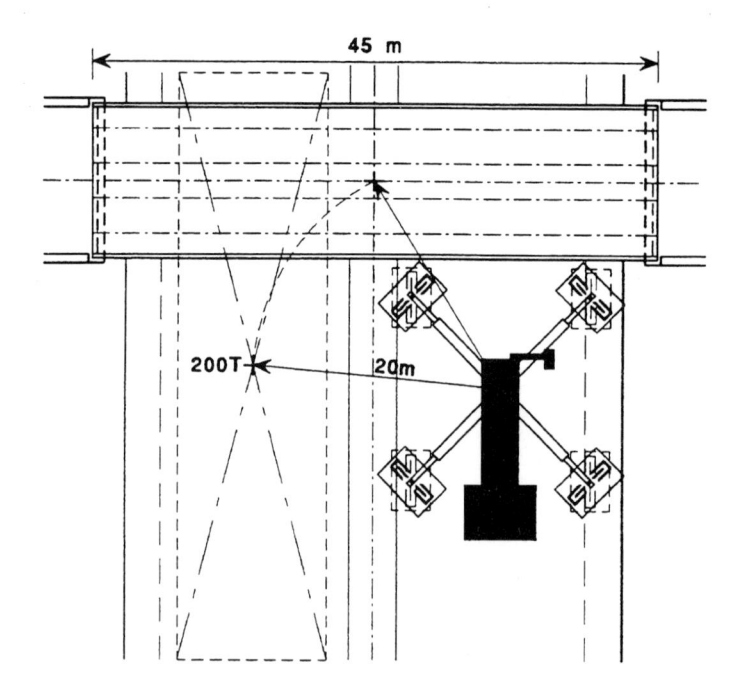

Figure 6 Erection of single span replacement motorway overbridge - Option 1

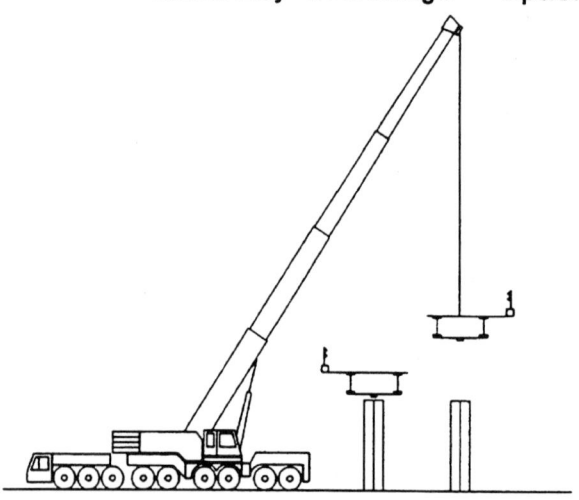

Figure 7 Erection of single span replacement motorway overbridge - Option 2

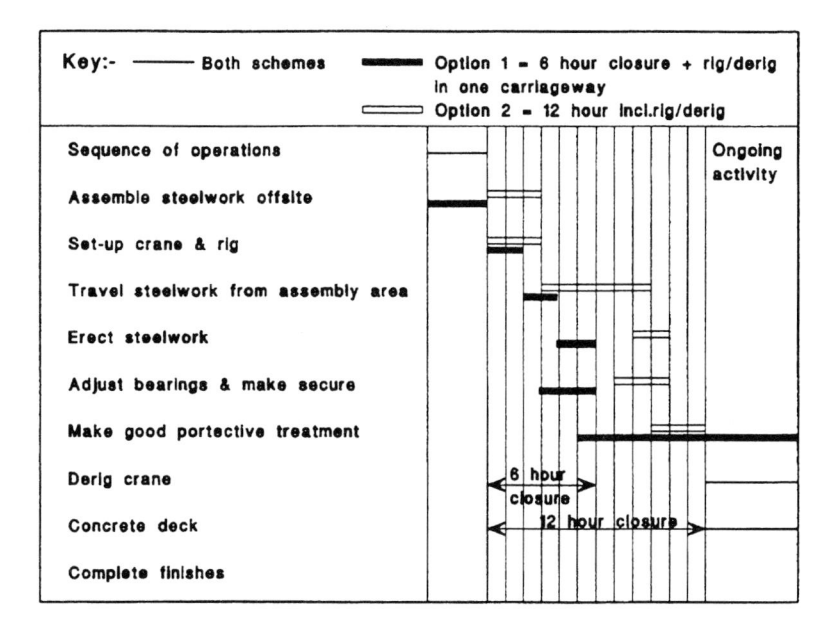

Option 1 Using single lift with 200 tonne @ 20m.
          Crane rigged/deriged in one carriageway (Figure 6)
Option 2 Multiple lifts using telescopic crane
          rigged during motorway closure (Figure 7)

Figure 8 Erection programmes for replacement motorway overbridges

PART 1: POLICY

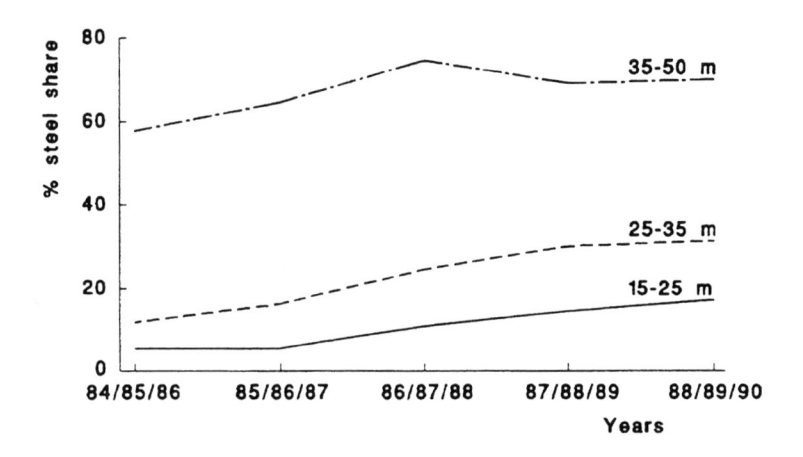

Figure 9 Market share for steel
in bridge construction

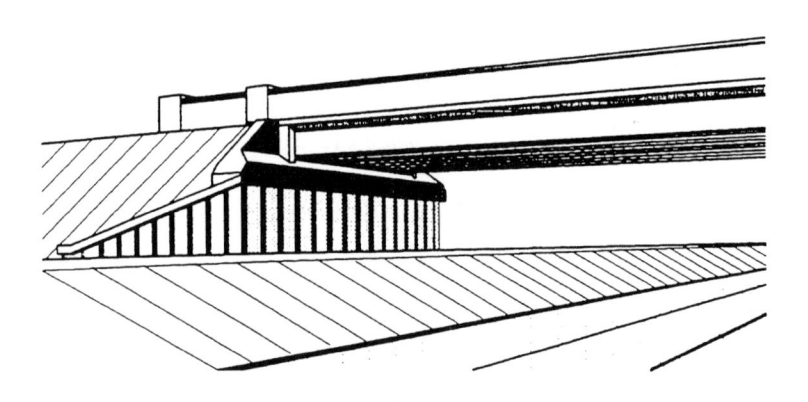

Figure 10 Steel substructure for
motorway overbridges

The concrete options

G. SOMERVILLE, Director of Engineering, British Cement Association

SYNOPSIS
1. Types of economic concrete section and methods of construction for bridges are reviewed generally. Durability performance (alkali-silica reaction and corrosion) is then considered, while indicating how these issues can now be dealt with, in design terms. Most of the paper is then devoted to the role of concrete in the context of the motorway widening programme; this covers the use of concrete generally, before outlining the options for both under and overbridges, involving precast and insitu concrete solutions. The key factors, unique to motorway widening roads, are identified, and it is demonstrated that there are suitable concrete options to cover the vast majority of situations.

INTRODUCTION
2. In the 1990's, the need to make bridges wider and longer is dominated by the motorway widening programme and this paper will concentrate on that. However, it is first necessary to look at the complete range of options for concrete bridges. Selection from that range will then become clearer - to meet the special needs for motorway widening, including traffic management, buildability and speed of construction (and subsequent management and maintenance in service).

3. The achievement of satisfactory performance in service is another broad issue which has to be addressed in presenting the concrete case. Much is now known about durability, based both on feedback and on research; this permits any required technical performance to be reliably achieved.

4. This paper first addressed these broad issues, before presenting the concrete options for motorway widening.

PART 1: POLICY

AVAILABLE FORMS OF CONCRETE SUPERSTRUCTURE

5. An excellent historical survey of developments in concrete bridges in Great Britain has been given by Sriskandan (1). This concentrated on the period from the Second World War to about 1986, and clearly showed the dominance of concrete, as the major road construction programme got underway in the late 1950s. For example, of the near 7,000 bridges built on motorways, truck roads and principal roads in England between 1965 and 1982, approximately 6,000 were in concrete, with over 50% of these being prestressed. Many of the spans were 30m or less, and a key factor here was the development of standard precast pre-tensioned beams. Standard bridge beams have been developed significantly since then, (2), (3), (4) with the Super-Y beams breaking the 40m span barrier.

6. Sriskandan (1) also reviewed trends in North America and Germany. In Germany, concrete was even more dominant, with prestressed concrete taking an ever increasing share (37% by number in 1984, 75% by deck area); these figures can be updated via a more recent paper (5), which shows 66% of all bridge decks being in prestressed concrete. In North America, Sriskandan noted a less buoyant picture, with prestressed concrete increasing from 12% in 1960 to 27% by 1973; however, a recent comprehensive survey (6) based on National Bridge Inventory (NBI) records show a somewhat different picture, as illustrated in Figure 1, with prestressed concrete accounting for half the bridges built in 1989. Detailed analysis of the data on which Figure 1 is based shows that the trend is mainly for spans up to about 45m.

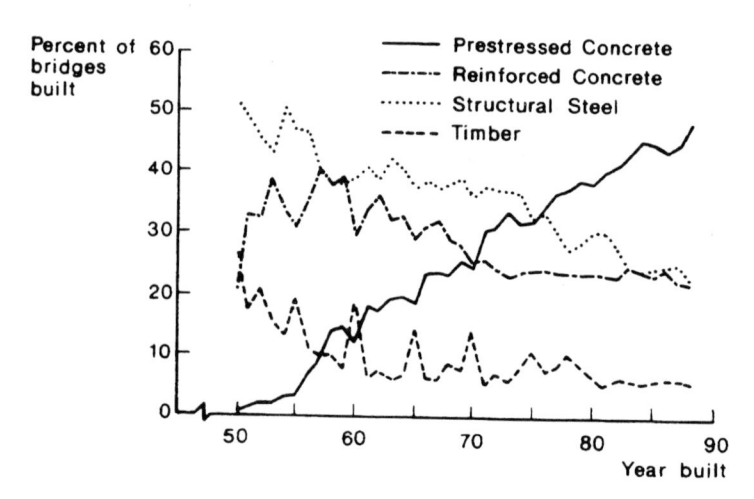

Figure 1. Percent of bridge types versus year built as of May 1989 NB (6)

7. Plainly, therefore, reinforced and prestressed concrete is in wide use generally for bridge decks. The next question is what form these decks can take. Lee (7) developed an idealised classification system, triggered by the development of box beams. This is shown in Figure 2, since the system is in fact general, leading outwards to orthotropic plates, either in the form of beam and slab construction, or of voided or solid slabs.

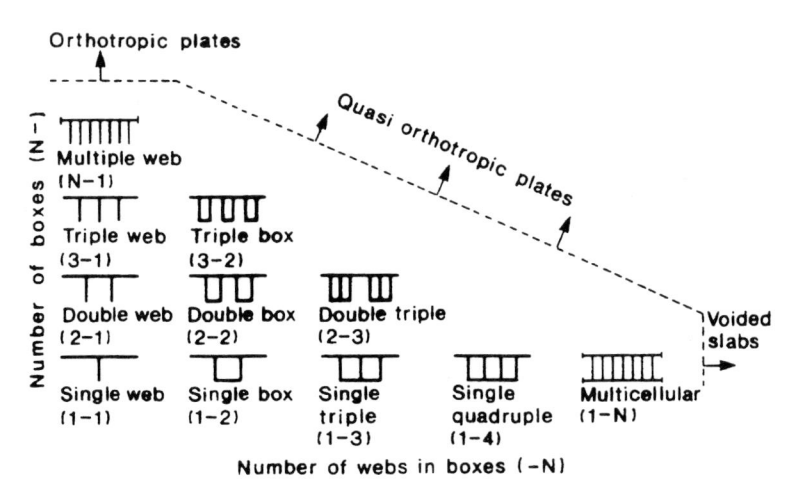

Figure 2. Classification system for concrete bridge cross-sections (after Lee (7))

8. In broad terms, Figure 2 indicates the type of cross-section available. However, this cannot be isolated from the method of construction. Figure 3, taken from reference 5, gives general guidance on the economic span ranges of different forms of construction; note that with some forms this economy is dependent on minimum lengths of bridge - in multi-span viaducts for example. Clearly, a number of these methods bracket the range of spans likely to be required for motorway widening; others may possibly be adapted to fit, provided all the other restraints, unique to widening projects, can be handled both in economic and practical terms.

PART 1: POLICY

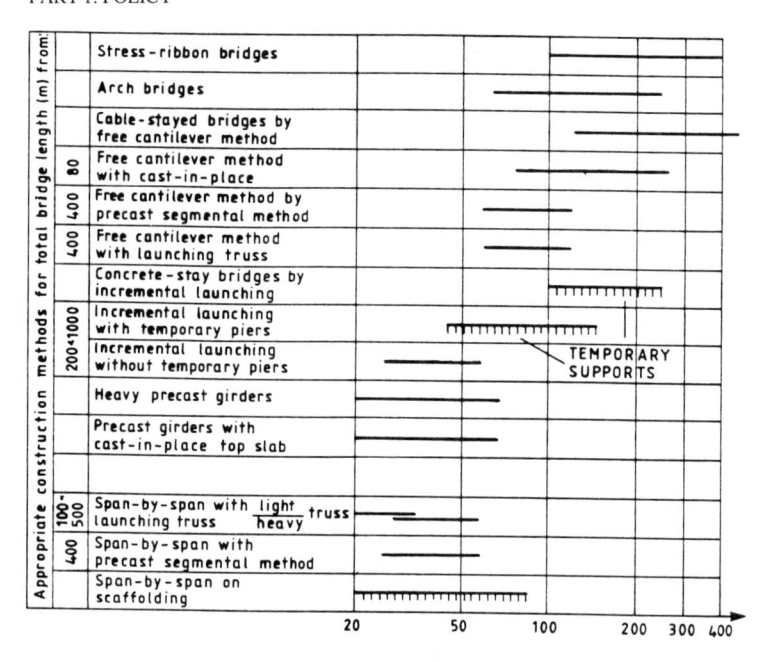

Figure 3. Limits of **modern** bridge construction methods for 20 to 400m spans (5)

9. In simple terms, Figure 2 represents the type of concrete section that can be called upon, and Figure 3 indicates how these can relate to different construction methods. Together they represent the pool from which concrete options may be drawn.

TECHNICAL PERFORMANCE
Durability - general

10. Concern over durability is perhaps the major issue in technical performance terms. Notionally, bridges are designed for a life of 120 years, although there is no clear relationship between that criterion and current wisdom and methods. It has also proved necessary to modify bridges on major routes for reasons of obsolescence, well before that notional period. Possibly, therefore, we need to re-think our basic objectives, so that the integration of initial design, maintenance and management leads to the fulfilment of real performance needs. This must come eventually from a life cycle costing approach.

11. In the meantime, we need to get more control over durability; greater durability may not be the economic solution, but rather a higher probability of achieving our prescribed objective, be it 10, 100 or 1000 years in service, with only routine maintenance and upgrading involved. Without having all the answers we are now well on the way, with regard to the 2 most

important issues for concrete - alkali-silica reaction and corrosion.

Alkali-silica reaction (ASR)

12. There are 2 factors here:-

(i) how to minimise the risk for new construction

(ii) an understanding and perspective of the structural significance of ASR in existing structures.

13. On (i), guidance is now embodied in the appropriate British Standard (8), based very much on the work of a Concrete Society Working Party (9). In effect, we now know how to handle this in design and specification terms and future amendments will involve refinement only, rather than a significant change in approach. On (ii), the second edition of the report by the Institution of Structural Engineers (10), when complemented by the BCA report on diagnosis (11), gives the necessary perspective. In particular, it distinguishes between structures where ASR is of little consequence, and those where further investigation is necessary - usually where the overall concept or the methods of detailing are particularly sensitive. While the Institution of Structural Engineers report does call for further research, this is aimed at refining the now established approach to assessment, i.e. the principles are clear, and we do have a proper perspective.

Corrosion
14. For bridges, chloride initiated corrosion dominates. For chlorides from external sources, we have to consider both sea water and de-icing salts. Of these, de-icing salts are more critical, since the moisture conditions - alternative wetting and drying - are more aggressive, and without benefit of the self-sealing of pores in sea water caused by the exchange of calcium and magnesium ions.

15. In considering the track record of concrete bridges with respect to corrosion (12), it is necessary to consider different types of construction, namely:

- reinforced concrete
- pre-tensioned prestressed concrete
- post-tensioned prestressed concrete

16. Reinforced concrete has been especially vulnerable to de-icing salts in bridge slabs, and especially in

PART 1: POLICY

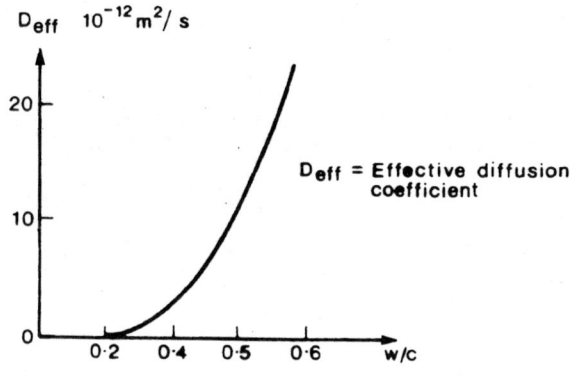

(a) Effective diffusion coefficient v water/cement ratio

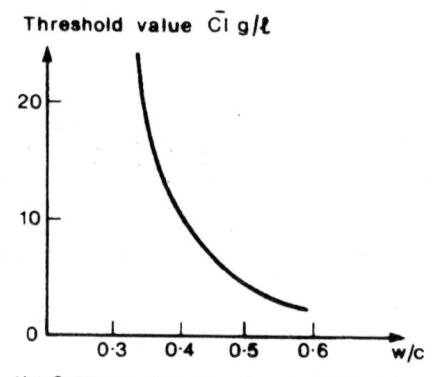

D_{eff} = Effective diffusion coefficient

(b) Critical chloride threshold level v water/cement ratio

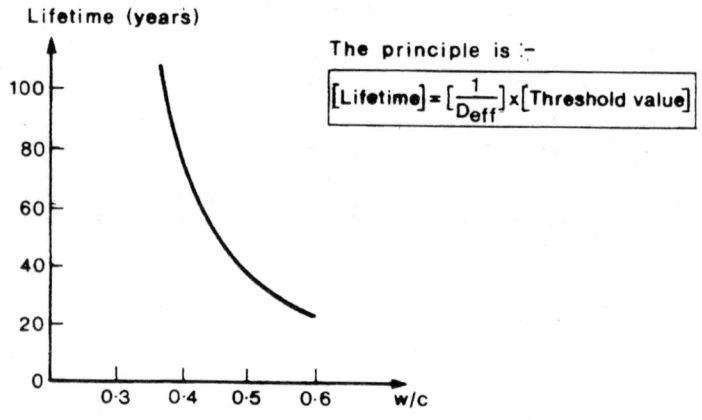

(c) Service life v water/cement ratio (as deduced from (a) & (b) above)

The principle is :-

$$[\text{Lifetime}] = \left[\frac{1}{D_{eff}}\right] \times [\text{Threshold value}]$$

Figure 4. The principles of good resistance to chloride-induced corrosion in concrete terms

end diaphragms and in sub-structures. This is due mainly to the lack (or failure) of waterproofing systems, and to deficiencies in joints, thus permitting chlorides in solution to directly attack the concrete, run down vertical faces and even pond. For soffits, there is the hazard of spray, and for piers, abutments and wingwalls a combination of splashing and spray. The situation has been made worse by sub-standard cover, but until recently, there has been no conscious design for de-icing salts other than specifying concrete quality and cover. As a result, maintenance and repair has become a growth industry, for all bridges in all countries where de-icing salts are used.

17. With precast pre-tensioned beams, we have a different story. There is no evidence in the literature of deterioration due to corrosion. Why should this be so? Firstly, the quality built into the method of manufacture means that the cover to all steel will have a high probability of being achieved. Secondly, the author has examined the production records of the major beam manufacturers in the UK (13); depending on the specification, concrete strength is in the range 60-80 N/mm^3, and the W/C ratio is less than 0.45 (frequently less than 0.4). If these facts are then related to the principles of good resistance to chloride induced corrosion, deduced from the literature (Figure 4), then the explanation is obvious; there is a steep curve, for both the effective diffusion coefficient and the critical threshold value for chloride level, which is dependent on W/C ratio and translates into an equally steep curve for the projected lifetime of the structural element. High strength concrete, as typically used for precast beams, is well placed on that curve; when linked to other basic requirements - described later - the potential to achieve any required technical performance becomes very real.

18. The high strength concrete argument also obtains for post-tensioned construction. However, the Achilles heel here has proven to be the grouting process. There is sufficient evidence to indicate that this has not always been done well, and that this is critical in the presence of chlorides. Particularly vulnerable areas are anchorages, and also joints where the jointing material has lacked the low permeability characteristics of the parent concrete. The problem is now being addressed with some urgency, by industry, and Government together; a good progress report is given by Raiss (14). An additional factor with post-tensioned construction is the use of external unbonded tendons, which is gaining favour

PART I: POLICY

partly because of ease of inspection and partly because of adaptability in terms of future upgrading of the structure. A good review has been provided by Lacroix (15), and it is understood (in September 1993) that a design method with a supporting specification will be published by the Department of Transport.

19. In summary, we now have a perspective in designing for corrosion resistance. The key feature is to get the basics right; the basics can be considered under 3 headings:-

1. <u>Overall design concept</u>. Avoid a vulnerable structure and sensitive details. Go for deck continuity, or integral bridges for lengths up to (say) 80m. Use mass concrete in substructures where appropriate. Develop a maintenance strategy.

2. <u>Detailing</u>. Waterproof the deck. Provide positive drainage and crossfalls. Consciously detail to deal with water in its various forms. Avoid joints, or detail them carefully.

3. <u>Concrete specifications and workmanship</u>. Carefully consider the quality of concrete required (see, e.g. Figure 4). Ensure good workmanship, with special regard to cover, curing, compaction, congestion (of reinforcement).

20. If these basics are right, then already we have several safeguards built in. If chloride concentration is judged to be severe, then there may be a need to consider alternative protection strategies. This should (and can) be done on a life-cycle cost basis, using predictive models to evaluate the relative technical performance of (say): coated reinforcement; surface membranes; special concrete layers; cathodic protection. Each of these techniques has its devotees as a panacea to avoid corrosion. It is the author's personal opinion that they are no substitute for getting the basics right, but should be considered as additions, in extreme cases. They also require further evaluation in cost-benefit terms; it is disturbing, for example, to see the current practice of silane treatment questioned in the literature (16).

MOTORWAY WIDENING - GENERAL REQUIREMENTS
21. The bridging solution adopted for a particular widening scheme depends on a wide range of factors. These include: traffic management (dominant); speed of construction and buildability; any provision for further widening and hence future adaptability; possible changes in functional requirements, e.g.

heavier loads, emergency lanes, etc; type and location of the motorway (urban, semi-rural, rural); land take and access; method of widening (parallel, symmetrical, asymmetrical); inspection and maintenance; safety; value for money; aesthetics; durability, including the provision of continuous or integral structures and attention to drainage; number of bridges/kilometre and whether involving on-line crossing or not; environmental issues in the broadest sense.

22. Even incomplete, this is a formidable list. The widening programme is still at a relatively early stage and the industry is on a steep learning curve. The situation is still fluid, with individual solutions being adopted for individual schemes. Moreover, the programme is proceeding at a pace which permits little time for considered alternatives or innovation, which perhaps the situation does require, to maximise benefits and value for money over a significant period. Certainly, consultation within the industry indicates variations in views on what constitutes the optimum solution in a particular case. A recent example (17) indicates some of the complexities - and the role of both concrete and steel bridges.

23. Any general generic approach to developing preferred and viable solutions is therefore beset with difficulties. In moving forward to proposing concrete options it is first necessary to draw-out those factors which are emerging as being most definite and dominant in this transient situation. These are:-

- the need to integrate bridge construction into the work programme for the whole scheme

- the dominant influence of traffic management and the costs assigned to delays

- the interaction of design and construction; buildability

- the trend towards bigger spans, uniquely for future flexibility reasons

- the trend towards continuity and integral bridges (for durability reasons), and hence the need to consider the total bridge

- the growing recognition of the functional role of bridges as part of the infrastructure and hence the need for more hands-on management with associated inspection, maintenance and upgrading.

PART 1: POLICY

24. Value for money over-rides all of these, of course - and there are other less tangible but important factors such as aesthetics and "green" issues where the expectations of the consumer are becoming better articulated.

25. Against that general background, the time has now come to establish the concrete options for the different applications.

THE USE OF CONCRETE - GENERAL

26. Superstructures tend to be the high profile part of bridge construction and certainly the area where choice of material is most significant. Before moving on to that, we need to recognise the role that concrete can play in other important areas.

27. Firstly, there are the traditional uses, in terms of foundations, sub- and retaining- structures. While alternatives may be considered on occasions, concrete will continue to dominate. However, there is scope for improvement, even for innovation, e.g.

- the adaptation of piers and abutments for greater efficiency in integral bridges

- wider use of precasting, both in sub and superstructures

- the use of permanent formwork including fair-faced or architectural finishes, or of permeable formwork for greater durability

- the use of high strength or mass concrete, in regions subjected to serve aggressive actions.

UNDERBRIDGES

28. Most underbridges involve spans of up to about 25m, perhaps increased for skew. Economically, this favours concrete, since there is a wide range of options available, both precast and insitu. The option selected will depend on a number of factors including:

- the construction used in the existing bridge
- the possible need to strengthen the existing structure to current standards
- the general condition of the bridge including the abutments
- the need to match span and appearance
- the phasing of the work as a whole, and access.

29. For the existing structure, necessary actions may range from doing nothing, via overspanning to total

replacement. However, in general new construction to current standards is first created on one or both sides of the existing bridge and traffic diverted on to that, prior to reconstructing or refurbishing the existing bridge.

30. For the span range indicated, some form of prefabricated unit will generally be the popular choice (Tables 1 and 2); this will be associated with one of the proven methods for achieving continuity (20) (21).

OVERBRIDGES
31. A realistic assessment of the potential for concrete bridges over widened motorways has been given by Thomas (18). A further strategy paper (19) deals with precast concrete solutions for symmetrical, asymmetrical and parallel widening, all of which require different spans.

32. The dominant factors affecting choice have been identified earlier in the paper. The required span is crucial in determining the viability of concrete options. In parallel widening on a rural motorway, for example, it is probably necessary to have a single span over the existing D3M, to avoid construction of piers in a live motorway; a slightly longer span will usually be adopted in this situation, for the reasons in the foregoing paper on policy, given to this conference by Boyes. It is understood that for trunk roads at least the possible need for future widening is based on traffic forecasts over 30 years; if the need is so demonstrated, then bridges are designed for this, provided that the additional cost is less than a prescribed percentage of projected modification or replacement costs for a smaller bridge. Decisions on exactly what to do are further complicated by:-

- the ability of the local road network to cope with the projected increased traffic;
- coping with the existing bridge where the new bridge is on line;
- assessing the benefits of reduced construction depth with shorter spans, versus the need for future flexibility;
- the ingenuity in devising alternative traffic management plans;
- contractor attitude in devising an economic management system for the whole scheme;
- the acceptable environmental footprint associated with the different options.

33. Against that background, it is not surprising that designers are investigating the merits of alternatives

PART 1: POLICY

Table 1 'Traditional' precast beams for spans up to 34m

"Y" BEAMS

SPAN LOADING 45 units HB Loading (incl 2·4kN/m² for finishes)

m	12	13	14	15	16	17	18	19	20	21	22	23	24	25	26	27	28	29	30	31
Y1																				
Y2																				
Y3																				
Y4																				
Y5																				
Y6																				
Y7																				
Y8																				

LEGEND ▬ Y Beams at 1000 centres

▒ Y Beams at 2000 centres

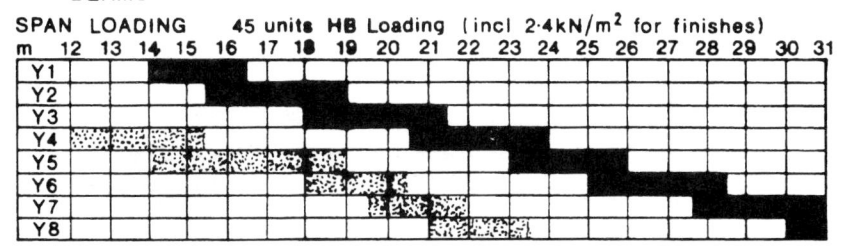

"SY" BEAMS

SPAN LOADING 45 units HB Loading (incl 2·4kN/m² for finishes)

m	24	25	26	27	28	29	30	31	32	33	34	35	36	37	38	39	40
SY1																	
SY2																	
SY3																	
SY4																	
SY5																	
SY6																	

LEGEND ▬ SY Beams at 1400 centres   Insitu Slab of 195

▨ SY Beams at 1600 centres   Insitu Slab of 195

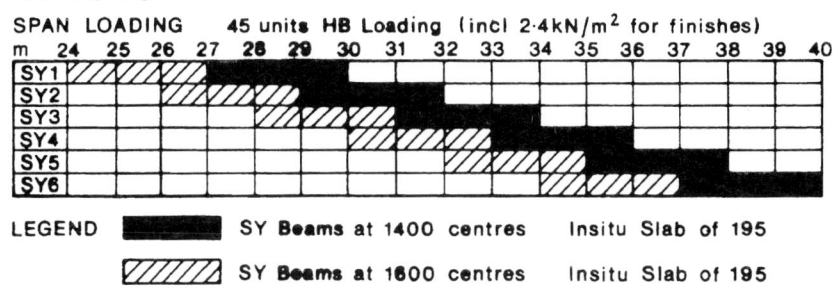

Table 2   The range of Y and SY beams for spans up to 40m (and above)

PART 1: POLICY

involving 2, 3 or 4 spans - and reaching different conclusions. What is clear is that it would be prudent not to place restrictions too close to the edge of a carriageway to maintain some measure of flexibility for the future. In spite of that, it is also clear that there are a number of concrete options that can cope with most of the spans involved - particularly for symmetric and asymmetrical widening, but also for many parallel widening cases - while also being viable in traffic management terms. Some of these are given below.

PRECAST CONCRETE OPTIONS
34. Standard precast pretensioned bridge beams really need no introduction. They have proved to be a popular economic solution for over 30 years, when the inverted T was first introduced. Since then, new and improved sections have been introduced, well researched and their track record in service proven - even in quite severe corrosive conditions. All the beams are fabricated by members of the Prestressed Concrete Association, and further details, including design recommendations, can be obtained from the PCA, at 60 Charles Street, Leicester.

35. The beam types currently available are indicated in Tables 1 and 2. For each type, and with 45 units of HB loading, an indication is given of the span range covered; in some cases for alternative beam spacings. These recommendations are for simply supported spans. The span range can be increased significantly by introducing continuity through composite construction. With good engineering design, span/depth ratios up to 30 have been obtained, thus improving aesthetics while reducing construction depth and hence the land-take in the approaches to the bridge.

36. The technical aspects of achieving continuity in this way has been well researched and design methods established over 30 years. Moreover, there is good feedback from performance in service, e.g. Pritchard and Smith (20); this together with American experience and practice, has led to proposals for fully integral bridges, e.g. Hambly (21). All of this is fully endorsed in the forthcoming DOT Advice Note on 'Design for Durability' (22).

37. We therefore have a proven solution, technically and economically, but in general terms. The next question is how this relates to the particular needs of motorway widening - as outlined earlier in this paper.

38. It is clear from detailed discussions with bridge engineers and from the work of a Task Group within the Concrete Bridge Development Group (18), (23) that the major problems are perceived as:-

(i) Transportation and erection, within the general phasing of the work.

(ii) Availability of beams to deal with the spans required by different methods of widening.

These will now be addressed.

Transportation and erection
39. Any prefabricated bridge beam, whether in concrete or steel, has to be transported and erected safely, with minimum disruption to traffic and in accordance with the phased work programme. Transportation of such units is governed by the Construction and Use Regulations. There are two basic issues:- the dimensions of the load and its weight.

40. For precast concrete bridge beams, the most critical dimension is usually length: if the length of trailer and load (excluding the towing vehicle) is greater than 27.4m, then a special order is required from the Department of Transport. In general, assurance will be required on the viability of the complete journey from precast yard to site, including the suitability of the route and the bridges. Reputable precasters are familiar with the system and the engineer can easily ascertain from them whether or not a particular beam can be carried to its destination, while routinely obtaining agreement in principle from the Department of Transport. As with any other form of exceptional delivery, the police must be informed (2 clear days notice) and agree to the timing. On the question of width, the limiting value, above which special provision has to be made, is 2.9m in the Construction and Use Regulations; the nature of these provisions depends on absolute width and on the timing of the journey.

41. The other main issue is weight; the version of the Construction and Use Regulations, seen by the author, defines an abnormal indivisible load at 38,000 kg. Above that weight, there are three categories and all the standard beams shown in Tables 1 and 2 fall into either category 1 (total weight not more than 46,000 kg) or category 2 (total weight not more than 80,000 kg). Within each category, clear indications are given of the number of axles required. Even the biggest unit shown in Table 2 - a Super Y beam of some 43m length and classed as category 2 for weight - can

PART 1: POLICY

be carried on a conventional articulated tractor and trailer with the trailer being steerable.

42. In short, an established system exists to permit the transportation of all the units shown in Tables 1 & 2, which is operated sympathetically by the Authorities, even giving broad commitments at the planning stage. This is familiar territory to all the precast manufacturers involved with bridge beams, who also know about transport availability and have yards conveniently close to motorway access points. In practice, such journeys will often be made overnight on a Friday or on a Saturday with a view to erection on a Saturday night. With proper organisation and management, it is perfectly feasible to erect a complete span overnight, with minimum disruption to traffic. Cranage for erection will obviously depend on the size and weight of the units and also on the length and height of the lift. A range of suitable cranes does exist; for Y and Super Y beams, telescopic cranes, with lifting capacities well in excess of 80,000 kg, and parked on hard standing, are perhaps the most appropriate in giving good rigging times for the erection of a complete span in a single night.

43. Current thinking in meeting the unique requirements for motorway widening - particularly for overbridges - puts a heavy emphasis on the use of some form of prefabricated unit, with the required span range tending to increase. The point being made in this section of the Paper is that proven precast solutions are available, as indicated in Tables 1 and 2, and that transportation and erection places no restriction on their use.

Availability of beams for different methods of widening

44. For underbridges, this is not an issue. For overbridges, it is. It is also relevant to the widening of trunk roads and to the building of new dual 3 or 4 lane roads.

45. For shorter spans, Table 1 is relevant. Table 2 indicates that the Super Y beams can cope with spans of at least 40m, depending on load intensity. Moreover, the ability to span greater distances is enhanced by the requirement for continuity (20). The 'insitu table' provided can, in theory, increase spans by 20% or more. In the reality of the motorway widening situation, however, it is unlikely that this extension will be greater than the depth of the beam, thus permitting temporary supports close to the final substructure - perhaps even by temporary brackets attached to that substructure.

46. In short, the potential is there waiting to be used. Long span precast beams have been used to provide overbridges over existing motorways, while maintaining traffic flow. Figure 5 shows the Walton Summit Bridge over the M61 under construction; the span of some 36m crosses a D3M.

Figure 5. Walton Summit Bridge under construction on the M61, using 36m span precast beams. (Photograph courtesy of C.V. Buchan Limited.)

INSITU CONCRETE OPTIONS
47. For longer-span underbridges, individual solutions can be developed from a consideration of Figures 2 and 3; it is a question of economic horses for courses, commensurate with the phasing of the work.

48. For overbridges relatively few conventional structures have actually been built (except for some 'dumb bell' crossings, or in special circumstances, e.g. the M1 (23) and M20 (17)). However, via consultation within the Concrete Bridge Development Group, the author knows of several schemes for a number of motorways, which are at the drawing board stage, usually involving prestressed box beam construction. In general, in-situ construction has come into its own in non-standard situations (heavy skews or odd geometry) or where a maintenance-free structure is required (17) or a premium put on aesthetics.

PART 1: POLICY

49. For more routine situations, the major issue is having sufficient vertical clearance to permit the use of temporary works clear of the traffic underneath. Conceivably, therefore, this is an area where innovation is most appropriate. Two promising possibilities come to mind, already in use overseas; these are:

1. <u>Incremental launching</u>. This has come a long way since its first use in the UK in the mid 1970s for the Shepherds House Bridge over the main Paddington to Reading railway line at Sonning (24). More recent applications have been documented (25) (26), demonstrating periods of 2 to 3 hours for movements up to 20 metres. The system was demonstrably economic for these applications and for motorway widening has the clear advantage of minimum disruption to traffic.

2. <u>Precast segmental construction</u>. As a technique, this is not new. However, its adaptation for motorway widening perhaps is novel. This has already been done for the Champfeuillet bridge, over the A48 motorway near Grenoble in France (27), by launching falsework in the form of two substantial steel beams, prior to sliding the match-cast units into place and stressing with unbonded tendons. Figure 6 shows the bridge under construction. An additional feature was that virtually the entire deck was precast in this way, leaving little more to do than the provision of handrails, services and surfacing.

Figure 6. Champfeuillet Bridge, Grenoble, under construction. 2 spans. Total length: 56.9m. (Photograph courtesy of Freyssinet International.)

CONCLUDING REMARKS

50. The detailed research necessary to produce this text has made it clear to the author that we are in a fluid situation, both in terms of defined need and perceived solution, for motorway widening - any individual case study, e.g. reference 17, makes that abundantly clear.

51. It has always been the case that the skill of the designer has lain in balancing the economics of the market place against required technical performance - perhaps while adding a dash of individuality, in terms of concept or detail, of choice of material, or of the different emphasis placed on individual factors such as aesthetics or structural form.

52. That situation has not changed. What has been added with the widening programme, however, is the special requirements for traffic management both during construction and maintenance in use, and the increased awareness of obsolescent and hence the need for greater hands-on management and appropriate performance in service. The first of these makes buildability a central part of the design. The second puts a strong emphasis on design concept and detail - in effect, on bridges as an economic and functional part of the infrastructure, while taking greater account of environmental issues.

53. In this Paper, an overall perspective is given of the various concrete options for bridges, in the context of upgrading motorways and primary routes, while taking the above factors into account. Detailed solutions are beyond the scope of the Paper, but they do exist; the objective is simply to make designers aware of the options and of where they can go for further information.

ACKNOWLEDGEMENTS

54. The author would like to thank many friends and colleagues who have given much of their time in helpful discussions while preparing this paper. He is particularly grateful to the Concrete Bridge Development Group, whose membership has provided much of the detailed information and positive support. It would be invidious to name names, but the individuals know who they are!!

PART 1: POLICY

REFERENCES

1. SRISKANDAN K. Prestressed concrete bridges in Great Britain: a historical survey. Proceedings of the Institution of Civil Engineers, Part 1, April 1988, Paper 9425, 269-303.
2. TAYLOR H.P.J., CLARK L.A. and BANKS C.C. The Y-beam: a replacement for the M-beam in beam and slab bridge decks. The Structural Engineer, Vol 68 No. 23, 4 December, 1990, 459-465.
3. REGAN P.E. Behaviour of precast prestressed Y-beams in shear, in torsion, and in negative bending. The Structural Engineer, Vol 68 No. 23, 4 December, 1990, 466-473.
4. HAMBLY E.C. and NICHOLSON B.A. Prestressed beam integral bridges. The Structural Engineer, Vol 68 No. 23, 474-481.
5. JUNGWIRTH D. and BONOMO R.J. Prestressed segmental bridges - the German experience. Concrete International, August 1992, 45-50. American Concrete Institute, Detroit, USA.
6. DUNKER K.F. and RABBAT B.G. Performance of highway bridges. Concrete International: Design and Construction, Vol 12 No. 8, August 1990, 40-42. American Concrete Institute, Detroit.
7. LEE D.J. The selection of box-beam arrangements in bridge design. Developments in Bridge Design and Construction. Crosby Lockwood & Son Ltd., London, 1971, 400-426.
8. BRITISH STANDARDS INSTITUTION, Concrete. Part 1. Guide to specifying concrete BSI, London, 1991. BS5328: Part 1: 1991, p.25.
9. CONCRETE SOCIETY. Alkali-silica reaction. Minimising the risk of damage to concrete. Guidance notes and model specification clauses, The Concrete Society, Slough, 1987, Technical Report 30, p.34.
10. INSTITUTION OF STRUCTURAL ENGINEERS. Structural effects of alkali-silica reaction - technical guidance on the appraisal of existing structures, Institute of Structural Engineers, London, July 1992, p.45.
11. BRITISH CEMENT ASSOCIATION. The diagnosis of alkali-silica reaction - report of a working party. BCA, Crowthorne, 1992, Report 45.042 (second edition) p.44.
12. DEPARTMENT OF TRANSPORT. The performance of concrete in bridges: a survey of 200 highway bridges. A report prepared by E.J. Wallbank, G. Maunsell and Partners, for the Department of Transport, Her Majesty's Stationery Office, London, April 1989, p.46.
13. PRESTRESSED CONCRETE ASSOCIATION. Private communication from individual members February/March 1993.
14. RAISS M.E. Durable post-tensioned concrete bridges. Concrete Vol 27 No. 3, May/June 1993, The Concrete Society, Slough, 15-18.

15. LACROIX R. External post-tensioning of structures. Joint seminar on developments in concrete bridges. The Institution of Structural Engineers/Concrete Bridge Development Group, 12 May 1993.
16. CABRERA J.G. and HASSAN K. The efficiency of organic polymer coatings on the durability properties of concrete. Proceedings of a European Colloquium on Construction Rehabilitation, Lyon, France, September 1992, 8-10.
17. HILL G.J. and JOHNSTONE S.P. Improvement of the M20 Maidstone by-pass junctions 5-8. Proceedings of the Institution of Civil Engineers, Civil Engineering, November 1993, 171-181.
18. THOMAS R.B. Scope for the use of concrete bridges over widened motorways. Proceedings of a joint seminar (Institution of Structural Engineers/Concrete Bridge Development Group) 12 May 1993. CBDG, Crowthorne, p.5.
19. PRESTRESSED CONCRETE ASSOCIATION. Motorway widening. Bridge replacement - the precast concrete strategy. February 1992. PCA, Leicester, p.11.
20. PRITCHARD B.P. and SMITH A.J. Investigation of methods of achieving continuity in composite concrete bridge decks. Contractor Report 247. Transport and Road Research Laboratory, Crowthorne, 1991.
21. HAMBLY E.C. and NICHOLSON B. Prestressed beam integral bridges, Prestressed Concrete Association, Leicester, 1991, p.29.
22. DEPARTMENT OF TRANSPORT. Design for durability. Departmental Advice Note BA/93. (In draft, April 1993).
23. CONCRETE BRIDGE DEVELOPMENT GROUP. Bridges for motorway widening. Report of a Task Group. Paper CBDG/TG5/1. CBDG, Crowthorne, March 1993.
24. BEST K.H., KINGSTON R.H. and WHATLEY M.J. Incremental launching at Shepherds House Bridge. Proceedings of the Institution of Civil Engineers, No. 64. February 1978, p.83.
25. ROWLEY F.N. Incremental launching bridges: UK practice and some foreign comparisons. The Structural Engineer Vol.71 No.7, 6 April 1993, 111-116.
26. IVERSEN N, FAULDS J.R. and ROWLEY F.N. Design and construction of the Dornoch Firth Bridge: design. Paper 10051. Proceedings of the Institution of Civil Engineers, Transportation 100. August 1993, 133-144.
27. ANON. The Champfeuillet Flyover - an innovative industrial process. The Freyssinet Magazine. December 1992, 10-11.

New ideas and innovations

A. C. G. HAYWARD, Cass Hayward & Partners

SYNOPSIS. Replacement highway bridges now need to meet the conflicting demands of longer span, heavier loading, accidental impact, solid parapets, minimal disruption during construction, maintenance and aesthetic appearance, all preferably within the same construction depth. This paper examines ideas and innovations which aim to meet these challenges within existing technology and construction skills. Innovations include the use of half-through construction for decks, monolithic piers, and the optimum use of steel and concrete proportionate to their advantages.

INTRODUCTION.
1. Motorway widening provides new challenges to engineers in designing bridges which can be easily built yet can meet criteria which are considerably more stringent than when the existing motorway network was built in green field conditions. Longer spans are needed and highway loading is significantly increased, both in terms of exceptional vehicles and the much higher proportion of commercial traffic. The incidence of "bridge bashing" demands supports and soffits of decks to resist accidental impact. At strategic sites high containment parapets are now demanded and these significantly affect overall design of bridge decks. Designers now need to consider the method of construction so that it causes minimal disruption to traffic. When motorways were built it was convenient use insitu concrete slabs and other simple forms, but fast-track programmes and the need to minimise disruption now demands prefabrication wherever possible. This puts extra demands upon design skills - allied to construction methodology and few organisations possess the combined skills. Hence the move towards design:construct contracts.
2. Maintenance and aesthetics are considerations which have only been brought to the attention of designers recently. It is desirable to minimise land take and to build replacement bridges on the same alignment and vertical profile. Thus designers are being forced to use much slimmer decks in proportion to their span, and this introduces

considerations of deflection and vibration. Other features of replacement bridges include the desirability of deleting piers within centre reserves to reduce collision hazard and the elimination of joints and bearings. A forthcoming Department of Transport advice note will encourage the use of jointless bridges, but experience with such bridges in the UK is comparatively limited.

3. Described in this paper are a number of ideas and innovations which are already in use or are within the scope of existing design and construction methods. Any attempt to describe esoteric forms which may well be developed in the future is not attempted. Solutions to the current challenges are needed now, that is over the next ten years. It is however correct to consider new materials and techniques being developed, such as the use of plastic composite materials, sky hooks or other wild fantasies. Such techniques generally require considerable development and proof of their suitability over a period.

MATERIALS

4. The main materials available to the bridge designer namely steel and concrete are likely to remain the first choice in the immediate future. Steel as available in the commonly used Fe510D1 grade has a yield of 355 N/m^2. There is some experience in the use of higher grades with yields up to 720 N/mm^2 and these could be useful in circumstances where flange plates would otherwise be excessively thick in bridges with low construction depth. Tapered steel plates (i.e. tapered in thickness and width) are now produced and have been used in France.

5. In concrete strengths up to 50 N/mm^2 insitu and 60 N/mm^2 precast are in regular use. There is advantage to be gained in use of increased strengths especially if current difficulties with post-tensional structures can be overcome. Other materials such as carbon and glass fibres are under active consideration by TRL for reinforcement of concrete. GRP and GRC have been used as permanent formwork to concrete, generally in a non-structural role. Recent completion of a footbridge at Aberfeldy in composite plastic materials is a valuable innovation. If an established track record can be achieved then these materials could well become acceptable and commercially viable for the structural elements of highway bridges.

Steel Orthotropic Bridge Decks

6. Welded steel orthotropic decks have been used since about 1950 in major suspension bridges, but also in moving bridges, roll-on roll-off linkspans and temporary flyovers etc. where weight savings are important. Intrinsic costs of a steel orthotropic deck are around three times higher than that of an equivalent reinforced concrete slab and this has discouraged wider use. However the economy of orthotropic decks has been largely governed by their use in the long span bridges. Longitudinal stringers have been of closed V or U trough form which introduce high demands on welding

PART 1: DESIGN

technology. Presence of heavy wheel loads on the deck causes local bending stresses between deck plate and stringer walls due to inherent torsional rigidity. High repetition loads from commercial vehicles has caused fatigue problems on a number of bridges in the UK and overseas. To an extent these have been overcome by research leading to better details (e.g. slot-through stringers) but high costs are still implied.

7. Use of open type stringers potentially removes virtually the construction and fatigue difficulties with orthotropic decks. Open type stringers can take the form of flats, angles, tees or bulb flats. It is necessary to design these stringers to carry individual wheel loads with little relief by transverse distribution so that the resulting steel deck is heavier than if closed sections are used. Open type stringers have been used in a number of roll-on roll-off linkspans and for example in the new Trafford Park Lifting Bridge due to open in 1994 shown in Figure 1. By use of open type stringers coupled with practical details for site joints it is considered that there is greater scope for the use of orthotropic steel plates in bridges over motorways to achieve increased span and greater width.

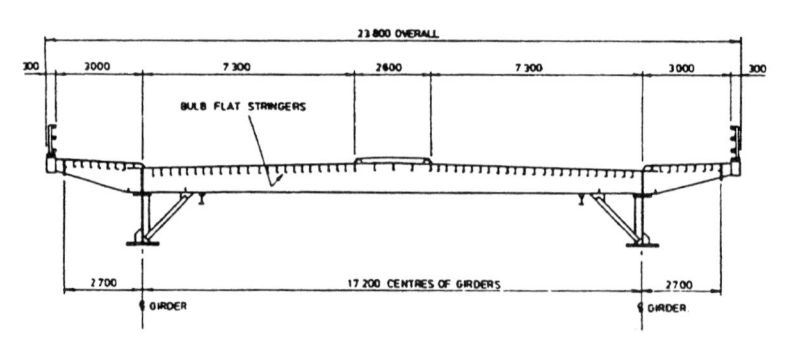

FIG 1. TRAFFORD PARK LIFT BRIDGE. BULB FLAT STRINGERS.

Lightweight Concrete

8. Lightweight concrete has been little used in UK bridgework. It deserves wider consideration because the savings in dead weight may well permit use of existing foundations for a replacement deck. A significant use was in the Friarton Bridge at Perth with main span of 174m constructed in 1980 (Ref. 1). Other UK use is limited to a precast 'M' beam bridge at Redesdale in 1969. The Arnhem bridge in Holland of 1986 is a notable example of 760m length (Ref. 2). A recent use is the Stracathro Interchange Bridge (Ref. 3) where lightweight concrete was used for the deck slab of a two-span composite bridge. Motivation for use of lightweight concrete was to reduce the weight for erection. Each of the preconcreted spans was erected with a mobile crane with telescopic jib under live traffic conditions.

Steel Fascias

9. Traditionally fascias on concrete or composite bridges consist of an extension of the slab cantilever about 0.5m wide and 0.5m high. Often the fascia is cast after completion of the deck slab and an undesirable construction joint results. Concrete fascias are vulnerable to degradation, staining, impact damage and poor workmanship.

10. A steel fascia using a rolled hollow section has advantages in that the parapet can be bolted directly and the fascia can be accurately aligned and levelled. It can be repainted as required together with the parapets. This form of fascia has been used on three recent highway bridges including the Tattershall Bridge at Lincoln (Ref. 4) as shown in Figure 2 and the Trafford Park Lifting Bridge referred to above.

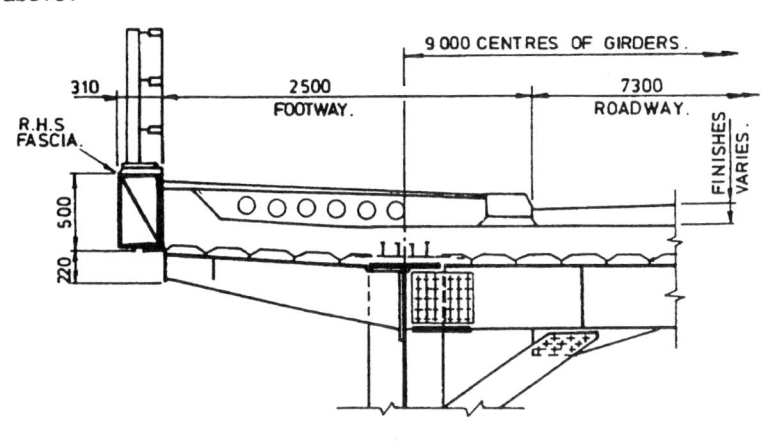

FIG 2 TATTERSHALL STEEL FASCIA.

BRIDGE FORMS
Deck Construction

11. Current practice is to use precast beams up to 30m span and steel beams to 50m and beyond, with span to depth ratio of around 20. Replacement bridges may well demand 30 or shallower. For example a 50m span replacement deck may call for a depth of 1.6m only. Unless half-through construction is adopted as described below then insitu prestressed slab or steel boxes will be the only solutions and the former will be problematic if falsework is to be avoided. Multiple steel boxes with steel orthotropic plate or steel shutter plate and insitu slab provide suitable solutions with the option of being rigid framed. Boxes are advantageous in not requiring bracing for soffit impact or erection stability. Figure 3 shows an example.

12. Recent developments include the use of corrugated webs for steel girders as used in France and currently being researched in the UK. They may be able to increase the economy of stiffened web construction.

PART 1: DESIGN

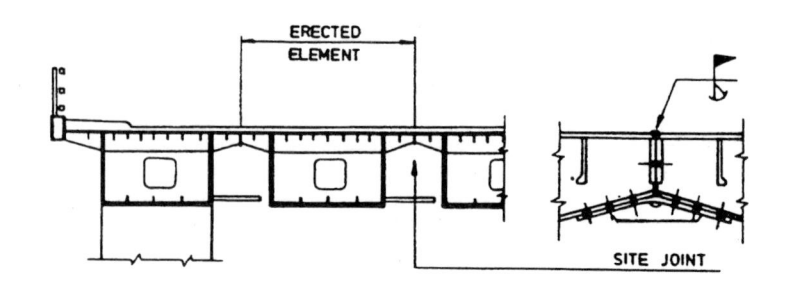

FIG 3. DECK TYPE BRIDGES. MULTIPLE BOX GIRDERS.

Half-through Construction
13. This form is used frequently for railway bridges where construction depth is at a premium, but less often for highway bridges. However, with the demands of longer span it is again being used. An obvious form is to use plate girders, supporting cross girders and concrete deck at the lower flange level as for a recent bridge on the inner ring road in Leeds. Steel cross girders are however inefficient. A feasible form of deck is to use standard inverted precast concrete T beams which span transversely and bear upon the girder bottom flanges. The two steel girders would be erected first and stabilised after which the beams are rapidly placed during a road closure, for concreting later. An important detail is the connection between the precast beams and steel web to ensure U-frame stability. Such a bridge form can be used for spans up to 50m, having a deck width of up to say 15m.

14. Upper limit of span is dictated by the stability of the top flange so that for larger spans then boxes would be used. The main girders conveniently form a high containment parapet and it is logical for footways to be mounted at top flange level. The footway slab would be made composite with the main girders. Services can be accommodated below the footway together with a runway beam for maintenance. Steel fascias would be used as referred to above and the external appearance made acceptable by presence of the footway cantilevers. Spans can be simply supported or continuous. Figure 4 shows cross sections.

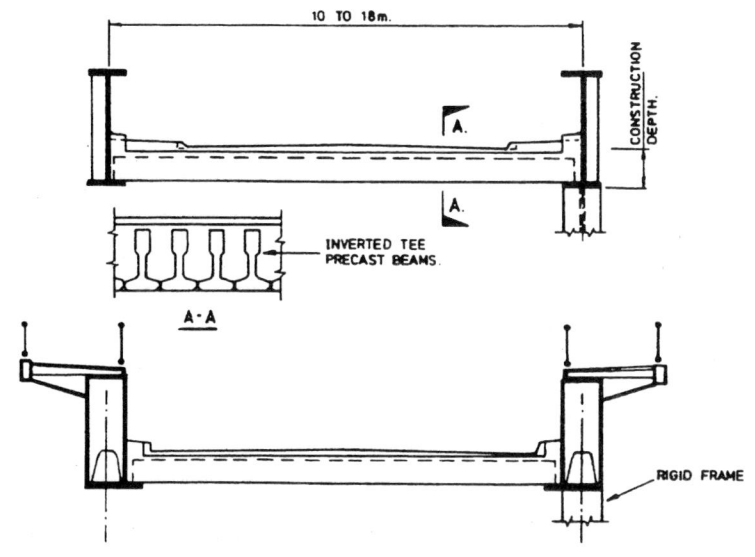

FIG. 4 HALF THROUGH BRIDGES

Precast Concrete Beams

15. Standardised precast beams placed side by side with concrete infill or top slab have been used for many years since the PCDG beams appeared in 1954. The standard M beam, inverted T beam, box beam, I beam are mainly displaced by the Y beam introduced in 1990, spanning up to 31m. A super Y beam (SY beam) has the capability of spanning up to 40m. These pretensioned beams are an excellent and durable form of construction. They have the disadvantage of limited span length and increased construction depth over steel beams which can more readily be connected at site for continuity. For precast beams then continuity can be achieved by insitu crossheads at supports and this has been successfully done on a number of projects.

Jointless Bridges

16. A DoT advice note is to be published which encourages jointless bridges for spans up to 70m in the interests of reduced maintenance, in that bearings and joints are eliminated. Jointless bridges have been used in the USA on a number of projects using both steel and concrete decks. Their use has been described by Hambly (Ref. 6). The advice note will be a challenge to engineers designing bridges up to 70m length. A number of details are possible at the deck ends but it seems essential to reintroduce running on slabs. A variant used by the author is to incorporate the abutment, curtain wall and wing walls as part of the deck, delete any expansion joints, and incorporate pin-type bearings beneath the deck which effectively props the abutments.

PART 1: DESIGN

Rigid Frames

17. Use of rigid frames is an extension of the concept of jointless bridges. There are a number of rigid frame type bridges in both steel and concrete where the deck height above ground is significant. Advantages are that bearings are eliminated and deck bending moments reduced so that rigid frames assist in achieving larger spans with short side spans. Steel deck beams were made rigid jointed with taper section columns on a bridge over rail station platforms at Bradford where insitu pier construction close to rail tracks would have been disruptive. Pier bearings will generally be necessary at ground level designed against accidental impact. See Figure 5.

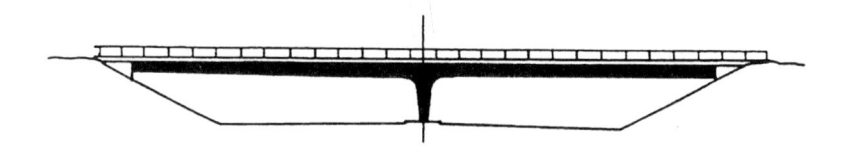

FIG. 5. RIGID FRAME.

Foundations

18. Foundations may contribute 50% of the cost of a bridge in poor ground. Recent developments include the use of reinforced earth and of steel sheet piling to form bridge abutments. These uses are likely to extend and are compatible with the concept of jointless bridges.

Trusses

19. For spans exceeding 50m with restricted construction depth then trusses may be appropriate. Recently they have been used for bridges carrying rail across motorway at Stepps, Glasgow, at Stockport (across M63, i.e. Bridge 70A) and over M5 widening at Avonmouth. A highway example occurs at Warrington (Ref. 6) which is of bowstring form using tubular members. This bridge has a span of 53m and a construction depth of only 1m and was almost the only solution because the road level was controlled by an adjacent existing roundabout. Truss bridges are promoted in the SCI publication on motorway widening (Ref. 7).

Cable Suspended Structures

20. Such structures of cable stayed or suspension type are used for long spans, say exceeding 200m, but have rarely been justifiable for shorter spans due to high costs including that of cables, their anchorages and the complexity of construction. In a situation of limited construction depth then a cable stayed bridge form could be appropriate for a motorway overbridge. The aesthetics needs careful treatment in urban areas. Cable stayed solutions should use the least number of cables necessary to support the deck in the

interests of reduced maintenance and better appearance.

Cradled Bridges

21. A concept for cradled bridges has been developed by Cass Hayward and Partners for converting existing superstructures by amending their mode of support to satisfy new loading and dimensional criteria (Ref. 8). The existing reinforced concrete deck is retained in its but is supported by two new primary girders located along each outer edge and spanning between new abutments constructed within the motorway embankments so as to provide an uninterrupted clearance of 59 metres and thus facilitate symmetrical widening of the motorway. The deck is slung using galvanised steel spiral strand ropes which are also used to raise the slab off its existing bearings. See Figure 6.

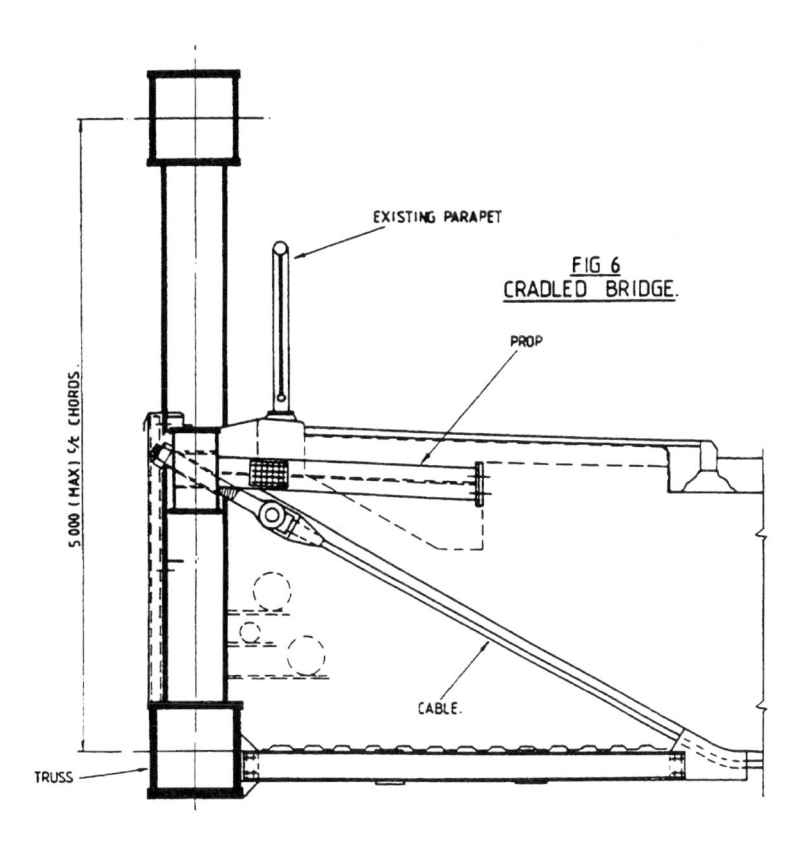

FIG 6
CRADLED BRIDGE.

PART 1: DESIGN

If additional soffit clearance is required the whole
structure is jacked up and the approach road alignment
adjusted accordingly. The primary girders are trusses up to
5m deep fabricated off-site and assembled to full length
using welded or bolted site connections prior to erection on
to the new bearings. Main chord members and end posts are
fabricated box members with web verticals and diagonals of
universal column rolled sections. Top chord compression
stability is provided by U-frame action developed between the
primary truss verticals and the existing deck concrete
utilising steel laterals introduced at the top and bottom of
the deck slab. Deep trimmer girders are provided between the
truss end posts which can be installed on new bearings set
inboard on the new abutment prior to main truss erection.
Fully rigid connections with truss and posts ensure stability
and stool connections into the deck soffit facilitate
longitudinal stability at the fixed end.

22. Principle of the cradling system is to utilise the
minimal depth afforded by catenary-type action of the
suspension system. Steel wire ropes are the traditional
medium but clearly some of the more advanced high strength
fibrous materials may be appropriate. The flexibility of the
ropes enables the variable profile of the deck soffit to be
assumed under load. Cables are provided at close centres
with steel block saddles at points of concentrated load
contact with the concrete. Saddles are shaped to an
underside fit with the concrete outline and profiled for a
suitable radiused interface with the cable. Cable sockets
are connected to threaded fork ends passing through the
longitudinal mid height beam of the truss providing anchorage
and allowing length adjustment of the ropes. Transverse
thrust from the suspension system is transferred through the
existing concrete deck.

Space Frame Decks

23. The concept has been proposed by Maunsells for using a
steel space frame form of deck with composite concrete slab,
which is soffit-clad using plastic composite material to
eliminate maintenance of the steelwork. The concept relies
on use of automated techniques in fabrication of the space
frame to reduce costs. At present the cost of enclosure
systems is high, but the incentive is to totally eliminate
future maintenance painting of steelwork.

CONCLUDING REMARKS

24. The above examples of ideas and innovations are merely
a selection of those which could be used or imagined, but are
considered to represent practical solutions for wider and
longer spans within available technology. Over the next ten
years it would appear appropriate to use steel concrete or
composite construction in proportions appropriate to their
economy. Both longer and wider spans are possible and by use
of these materials in higher strength their efficiency can be
improved. The challenge to design efficient structures which
can be built easily, maintained infrequently and are pleasing

to look at will continue to test the ingenuity of bridge engineers.

REFERENCES
1. KERENSKY, Dr. O.A. The design and construction of Friarton Bridge. The Structural Engineer. December 1980. Volume 58A. Number 12.
2. Arnhem Bridge test run for Lytag mix. New Civil Engineer. 13 February 1986.
3. HAYWARD, A.C.G. A Bridge in a Day. Steel Construction Today. Volume 5. Number 1. January 1991.
4. HAYWARD, Alan. Tattershall Bridge, Lincolnshire. New Steel Construction. Volume 1. Number 5. August 1993.
5. HAMBLY, E.C. & NICHOLSON, B.A. Prestressed beam integral bridges. The Structural Engineer. Volume 68. Number 4. December 1990.
6. HAYWARD, Alan. Tubular members in bridges. Tubular Structures, Bridges. Publication ISSN 0041 3909. May 1990.
7. Replacement Steel Bridges for motorway widening. SCI/BCSA/British Steel publication 204. 1992.
8. HAYWARD, A.C.G. & SADLER, N.L. Cradled bridges and the reuse of existing decks. Symposium in Bridges - Extending the lifespan. Leamington. May 1992. Construction Marketing Limited.

Widening bridges for the M1 and M6 motorways

A. M. PAUL, Director, and R. S. COOKE, Associate, Ove Arup & Partners

---

1. In 1990 the Department of Transport announced that it would undertake a programme to extend and increase the capacity of the nation's road network. The central part of this programme included plans to widen several existing motorways which are heavily trafficked and frequently congested. The M1 (Junctions 15-19) and M6 (Junctions 11-16) which are considered in this paper, are two of the oldest sections of motorway, originating from the late 1950's and early 1960's, respectively.

2. The viability of widening these routes depends on numerous factors, including traffic, the highway itself, bridges, ground conditions and the environment. Natural constraints are also created by existing towns, roads, railways, rivers or other geographic features.

3. Bridges are a major element of these schemes. Their modification or replacement may dominate the sequence of work and comprise a substantial cost. This paper describes some of the features which are being developed for the parallel widening of these two sections of motorway.

SCHEME REQUIREMENTS

4. The M1 and M6 motorways considered here have dual 3-lane carriageways with hardshoulders and central reserve. There are several ways to increase traffic capacity by adding further lanes. These methods have been described in detail by others (Ref. 1,2) and include symmetric, asymmetric and parallel layouts. These layouts are shown in Figure 1. There is nothing inherently new in these concepts. Long sections of the M5 motorway in Hereford and Worcestershire have already been widened from 2 to 3 lanes by each of these methods. The specific choice of widening layout depends on numerous factors. For the M1 and M6 motorways parallel widening is favoured because these sections are largely in rural areas with few urban constraints. Widening in the vicinity of built up areas can be implemented away from housing whilst preserving any existing screening or noise protection adjacent to the motorway. Elsewhere the parallel carriageway is built so that traffic management requirements are simplified and the number of crossovers, positions where the widening changes sides, are minimised.

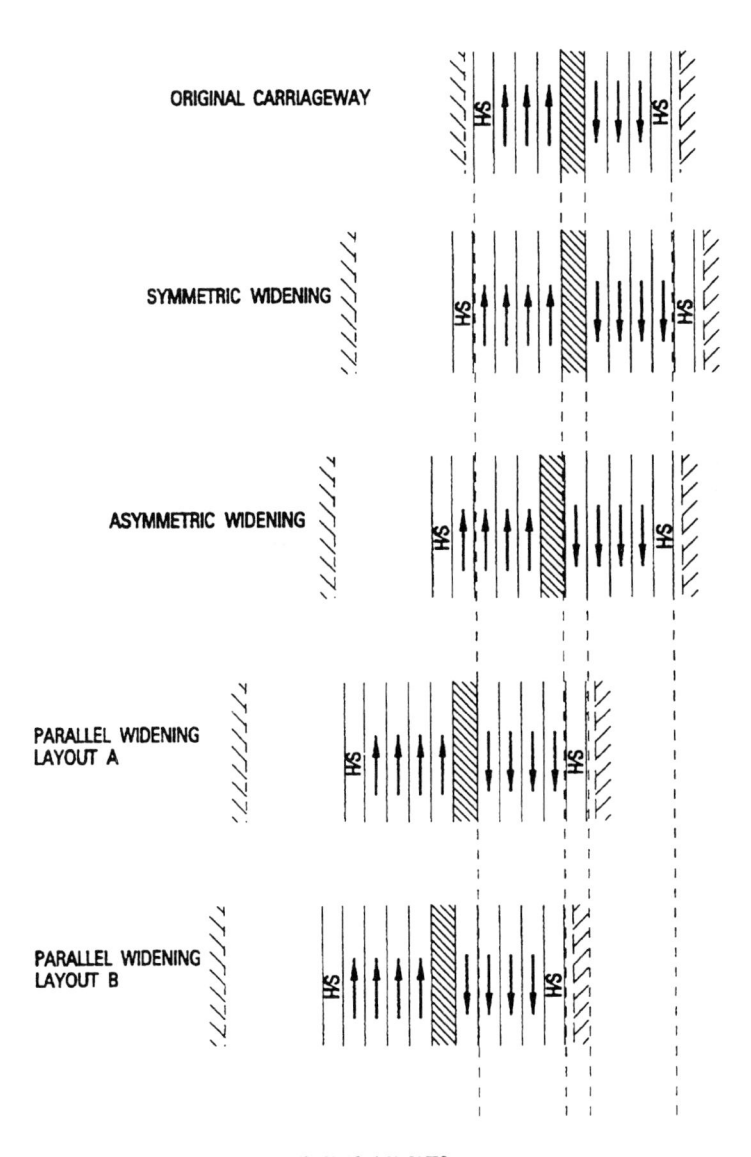

FIGURE 1 MOTORWAY WIDENING LAYOUTS

5. The principal difficulties that arise for these schemes are the high traffic flows, wider widths of carriageway and increased awareness regarding safety and overall disruption. As a result the bridge requirements are considerably more onerous, limiting some conventional forms of construction and encouraging the use of new methods.

PART 1: DESIGN

EXISTING BRIDGES

6. The existing bridges consist of a variety of constructions. On the M1, many overbridges and underbridges are built in classic Sir Owen Williams style. (Ref. 3). There are also some more unusual underbridges over rivers and a canal. The majority of bridges are made of precast and insitu reinforced concrete or mass concrete.

7. The M6 motorway partially consists of more Sir Owen Williams bridges South of Junction 13 and structures designed by Staffordshire County Council to the north. A greater variety of structures exist which also include composite bridge decks with precast, post-tensioned beams and concrete slabs with encased steel beam sections.

8. The condition of these bridges varies across the full range that might be expected for bridges approaching 35-40 years of age. Extensive maintenance is necessary in some cases where joints have failed or reinforcement with low cover is corroding. Conversely bridges built using mass concrete, often without expansion joints are performing very well with few signs of deterioration.

9. Appraisal of the existing bridges has mainly concentrated on the underbridges as it was quickly apparent that most overbridges would need replacement to accommodate 4 lane carriageways. Structural assessments of the existing bridges have been undertaken. These assessments have considered the load carrying capacities of bridge decks, substructures and foundations for all underbridges as well as the effects of impact loading from vehicles on overbridge piers. The assessments have shown that some bridges are below current loading requirements for permanent and live loading effects. These existing bridges will require either strengthening or replacement as a part of the scheme proposals. The requirements of the DOT's 15 year Rehabilitation Programme (Ref. 4) have also been considered.

10. Whilst these appraisals form a very significant proportion of the work needed for the scheme design, the main emphasis of the work concerns the new structures needed for the widened motorway.

NEW STRUCTURES

11. The development of a new scheme allows a number of options to be considered. The benefits of past experience and new construction methods are ideally combined to achieve buildable, cost effective solutions. These large schemes encourage the use of consistency for groups of structures. This gives an identity for the car driver and repetitiveness for the contractor and designer, leading to overall economy. However some bridge sites

present the opportunity to provide slightly more elaborate solutions, so called landmark structures. The following sections describe the forms of structure being considered.

12. <u>Motorway Layout</u> For both the M1 and M6 schemes a choice of parallel widening layout was considered. Layout A involves positioning one of the new 4 lane carriageways over the existing central reserve, whereas in Layout B the new carriageway is shifted away from the central reserve. These two arrangements are shown in Figure 1. Differing benefits arise from these two options. For example Layout A favours the existing underbridges because existing parapets can be maintained during construction and the edge of the existing bridge can become a natural position for a joint separating it from the new bridge needed to support the parallel carriageway. New overbridges are also slightly shorter with an approximate square span of 39-40m needed over the existing motorway. However there are drawbacks associated with the traffic management arrangements needed when the old carriageways are modified. Layout B allows the retention of the existing central reserve during construction. Thus the safety fencing and any existing lighting can be maintained, providing traffic safety benefits. Traffic management is also simplified and the benefits of this arrangement can be calculated using QUADRO and COBA because shorter lane restriction periods are predicted during which traffic is delayed. The additional costs associated with extra construction width and longer or wider bridges are more than balanced by the benefits achieved.

<u>Overbridges</u>

13. For a parallel widened motorway the existing overbridges do not always provide sufficient spans or length to accept the revised carriageway layout. Hence total replacement is the only option. Unless suitable diversions are available on the side road it is usually necessary to build and complete the new overbridge prior to removal of the existing structure. Consequently the span arrangement of the new bridge must suit both the existing and proposed motorway layouts. For on-line replacements, temporary parallel bridges are needed instead.

14. <u>Spans</u> The overbridge spans are chosen to satisfy the existing and proposed carriageway layouts. The absolute minimum spans are shown in Figure 2, which indicates that square spans of 43 and 25m are required. However it was considered that such imbalanced spans could lead to undesirable visual results in certain locations. Effects such as skew, curvature, gradient, construction depth and headroom must all be considered. A second factor is also important. Traffic estimates for these primary routes suggest that further widening may need to be considered in the future and it is important to provide bridges with flexibility to meet this. This is relevant bearing in mind that bridges have a 120 year design life. It is not possible to be certain how the motorways may need to be modified within this timescale, but increases in spans for the overbridges will allow

PART 1: DESIGN

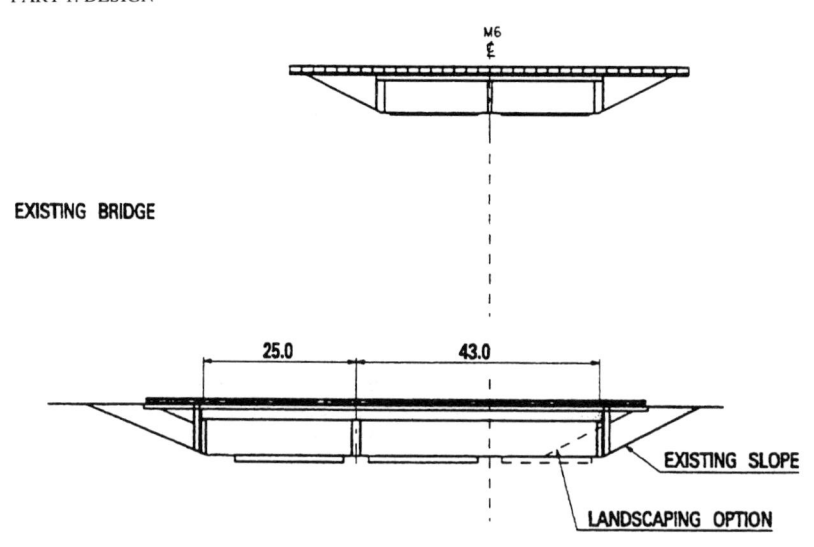

EXISTING BRIDGE

NEW BRIDGE

FIGURE 2 ASYMMETRICAL BRIDGE

carriageway layouts to be changed without having to replace the bridges within their normal design life if an obstruction is caused by piers or abutments. Thus the proposed layouts shown in Figure 3 are being considered, using spans of 43m and 37m.

15. <u>Appearance</u> The overall appearance of a bridge depends on many factors. The bridge is part of a landscape and not just a thing in itself. Bridges in a rural setting should generally be as inconspicuous as possible, simple and consistent in general character and detailing. A group of bridges on a length of motorway should be considered as a whole as well as individually. Both unity and variety are desirable and bridges should clearly relate to each other, but also be recognisable as individuals. General aesthetic considerations apply such as proportion of spans to each other, relationship of solids and voids and the deck depth to clear height over the road. For rural motorways there is a preference for open end spans with inconspicuous bankseats rather than solid abutments which are more obtrusive.

16. Bridges for motorway widening give rise to a number of problems. Over long spans the depth of the deck is likely to be large in relation to the clear height above the motorways. This will cause the opening to appear as a thin slot with a heavy deck looming over it. This can be mitigated in the following ways:-

- by increasing the clearance height
- by reducing the structural depth
- by using a variable depth beam

EXISTING BRIDGE

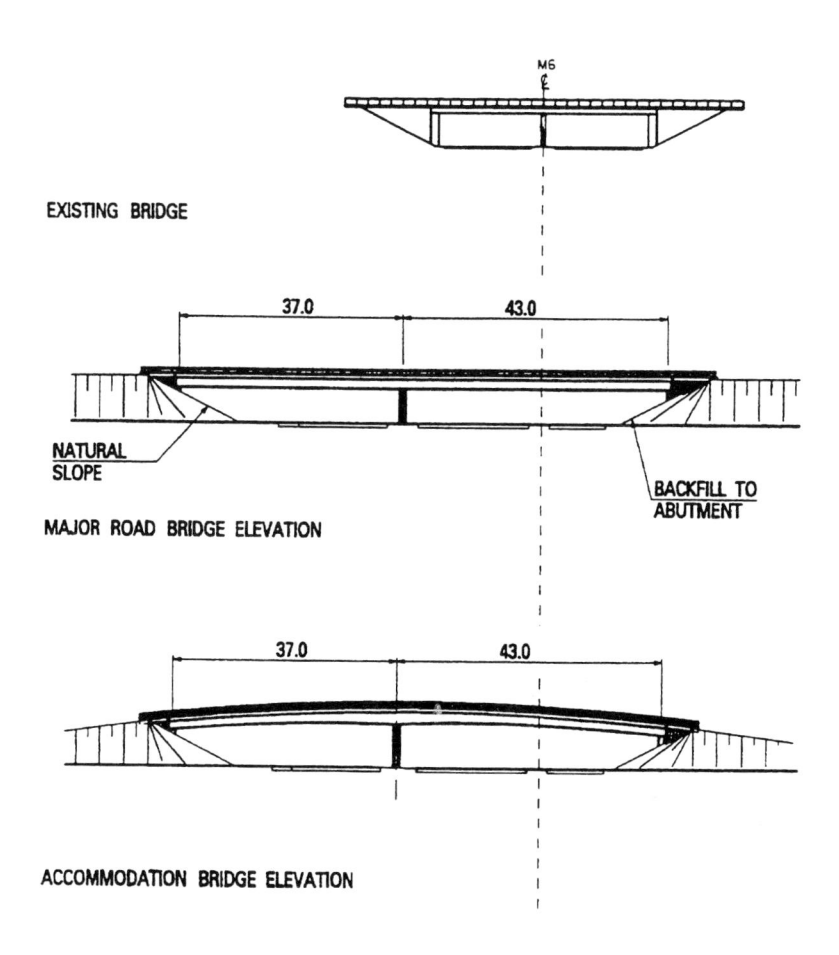

MAJOR ROAD BRIDGE ELEVATION

NATURAL SLOPE

BACKFILL TO ABUTMENT

ACCOMMODATION BRIDGE ELEVATION

FIGURE 3 PROPOSED OVERBRIDGES

- by using a cross-section which makes the real depth impossible to read; trapezoidal and curved sections have been used successfully.
- by emphasising a well-lit fascia beam over a structure which is cast in shadow by a large edge cantilever; the use of a dark colour for the main structure can have similar effects, but some cantilever is still necessary.

17. The first three are real solutions which alter the ratio of solid to void. The last two are visual effects which can minimise the appearance of depth, although they do not work in all lighting conditions because they do not change the silhouette.

PART 1: DESIGN

18. <u>Visualisation and Models</u> In order to investigate the general form and proportions of typical structures computer visualisations and models have been used. The computer views can be generated with added colour and shading to test the concept of an individual detail or complete arrangement. Such methods are generally satisfactory but still limit the eyepoint to a selected number of viewing positions. Models, however, provide an excellent medium to test the appearance and proportions of a structure in three dimensions. The value of making models should not be overlooked at the preliminary design stage as a means of conveying ideas or proposals.

19. <u>Form of Construction</u> Experience has shown that the decks of new overbridges must be built as quickly as possible without lengthy disruption, so prefabricated sections are preferred. A comparison between prefabricated steel and concrete sections was made as shown in Figure 4. This shows that steel sections are favoured in this case, primarily due to the need for spans of at least 43m on one side. This would increase for skew bridges, thus ruling out precast concrete beams such as the SY (Super Y). Other benefits related to reduced construction depth, flexibility to satisfy curved alignments (horizontally and vertically) and overall weight are achieved using steel. However there are also some disadvantages or problems which need to be considered.

COMPARISON OF CONSTRUCTION OPTIONS FOR OVERBRIDGES

PREFABRICATED ELEMENTS	STEEL COMPOSITE	PRECAST CONCRETE BEAM
Span	20-60m	45 Max
Construction Depth at 40m	1.6m	2.2m
Span/Depth Ratio	25:1	18:1
Skew	Unlimited	≃45° (17° for 45m span)
Collision	Vulnerable	Vulnerable
Curvature?		
Vertical	Yes	No
Horizontal	Yes	No (Limited effect)
Continuity	Yes	Yes
Weight	Low (25T)	High (95T each)
Maintenance	Painting	Silane

Figure 4

20. A typical cross-section for a standard overbridge is shown in Figure 5. The main points to note are:-

(a) Cross-bracing is needed both to support the beam pairs during erection and to assist the resistance of the steel beams to vehicle impact. Increasing the headroom under the bridge to more than 5.7m can be economically justified and reduces the frequency and size of the crossbraces, although they are still needed at supports.

(b) The service bay is positioned in the composite slab, so that maintenance access can be achieved from above and so that the motorway is largely protected from any ruptured services. To minimise depth, cover slabs are not being considered, but hardened verges will be required.

(c) The parapet edge beam and cantilever is often difficult to construct over live traffic. Extensive falsework is needed for insitu casting of the edge beam concrete. Casting the slab and edge prior to lifting the beams may be viable, but care is needed to ensure that a smooth accurate line is obtained prior to application of other permanent loads or adjacent spans. Precast solutions may also be feasible, but frequent joints need to be detailed in a way so that potential water leakage and staining is avoided.

One solution to simplify this construction is to reduce the cantilever overhang. However, this is considered undesirable because the apparent depth of the bridge is exaggerated and the exposed elevation of the steel beams may show dirt and water staining. The use of a dark paint colour combined with shadow cast by the cantilever combine to minimise the impact of the beams and reduce apparent depth as discussed above.

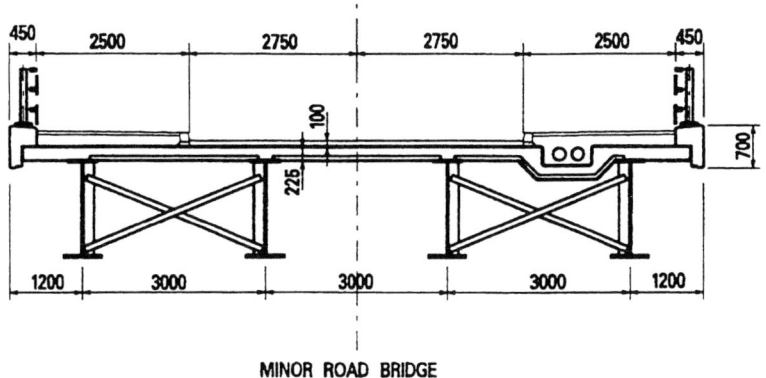

MINOR ROAD BRIDGE

FIGURE 5 TYPICAL OVERBRIDGE CROSS-SECTION

PART 1: DESIGN

21. <u>Substructures</u> The abutments and piers for the overbridges tend to be less obtrusive than the decks. Existing motorway overbridges viewed in a mature landscape are generally recognisable due to the prominence of the deck rather than the substructures. These elements must therefore be designed to be functional in the first instance. The use of bankseats or skeleton abutments at the top of the side slope to give an open appearance means that there is little to see in a normal situation. The need to have easy access to joints and bearings is often helped by having a slope leading to the end of the bridge.

22. For bridges with 43/37m spans a future modification to the motorway layout may seriously affect the abutments, whereas the pier would probably be unaffected. Apparent symmetry of bridge appearance is desirable for any carriageway layout, although this is not always possible. Consequently the arrangement adopted after initial construction allows for some soil backfilling around the abutment adjacent to the existing motorway to give a similar appearance to the opposite abutment (see Figure 3). If the carriageway is further widened, so that the original pavement is reused, then this side slope must be removed. The abutment will be exposed over its full height and additional retaining walls needed. The isometric detail of the abutment is shown in Figure 6. The main point to note is that allowance should be made in the design for total removal of the frontfill. The addition of retaining walls constructed approximately in line with the abutment wall can be achieved without any disruption to traffic over or under the bridge. Temporary sheet piling will be needed to allow for construction of the foundations. Figure 7 shows the final arrangement. Only minimal account needs to be taken of these additional walls which would be free standing on their own. Thus no additional expense is needed when the main bridge is built to allow for these future changes, whose precise details cannot always be determined at this stage. It is anticipated that the main abutment wall should be protected with plastic sheet while it is buried and may need to be grit blasted upon exposure to obtain a consistent concrete finish.

<u>Underbridges</u>

23. A variety of existing underbridges varying in size from small drainage culverts and accommodation structures, to long multi-span viaducts exist on the schemes.

24. <u>Motorway Layout</u> For parallel widened construction, the new carriageway needs to be supported on a new bridge usually built directly in line with the existing bridge and providing similar clearances. Separate construction with a joint to the old structure positioned within the central reserve will allow some flexibility for differential settlement. However the use of a widening Layout B means that the existing bridge must be extended by about 3.5m to ensure this joint is not located under

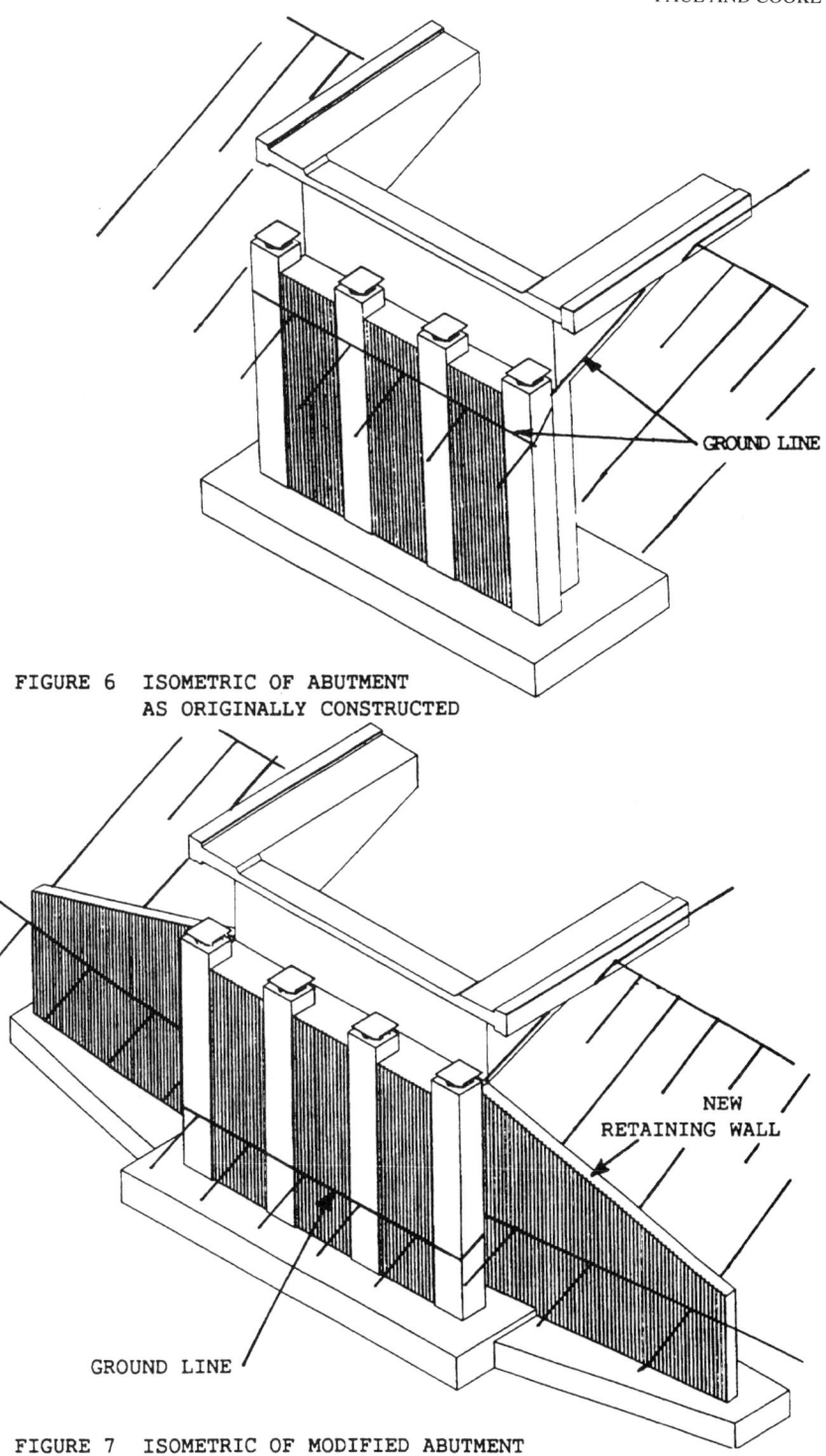

FIGURE 6   ISOMETRIC OF ABUTMENT
AS ORIGINALLY CONSTRUCTED

FIGURE 7   ISOMETRIC OF MODIFIED ABUTMENT

PART 1: DESIGN

the running surface. This is the major disadvantage with using Layout B. However Layout A can be equally problematic for bridges with central reserve lightwells because difficult and costly modifications are needed on several structures in a location where working space is limited. The extension will normally reflect the existing construction and articulation. The means of building these extensions must be carefully considered in traffic management terms. Departmental Standard IM5 restricts the proximity of running traffic for up to 7 days, so that vibrations and deck deflections are minimised on curing concrete. This restriction is not a practical proposition for motorways which are heavily trafficked. Research and investigation, principally in North America, indicates that some vibration can actually be beneficial to curing concrete. (Ref. 5). However each situation must be considered on its own merits and the actual method of construction carefully vetted once the scheme is being built.

25. <u>Railway Crossings</u>   The construction of widened motorways over existing railways, particularly electrified main lines, can pose problems. In many cases existing spans provide only nominal clearance over the railways. New regulations (Ref. 6) require increased lateral clearances between rail lines and closest obstruction, eg, foundation or abutment walls. For situations with Layout B it may therefore be necessary to extend an abutment wall parallel to the railway at the same offset as the existing sections. There is often limited scope for operating restrictions or possessions on railways and hence extensive works on these bridges may be expected in some cases. A change in motorway widening layout, for example to Layout A could be viable, but might incur hidden costs in the form of traffic management and delays which outweigh the alternative structural and BR costs.

26. <u>Appearance</u>   Underbridges and viaducts are much less noticeable to the motorway user. The need for consistent appearance is less pronounced, but underbridges can experience must closer scrutiny from users, particularly pedestrians. In an urban area the use of brick or textured concrete facing may be appropriate. Development of these proposals is not sufficiently advanced to describe at this stage. Also the use of supplementary features much as noise barriers can significantly effect the appearance of a bridge and needs to be considered sooner rather than later.

27. <u>15 Year Rehabilitation Programme</u>   Motorway widening provides an opportunity to undertake major maintenance and strengthening of existing underbridges which might not be viable in other circumstances. The appraisal of current strength, durability and deficiencies must be considered in the context that maintenance or repair, which may be needed in the foreseeable future, should also be carried out whilst the motorway is being modified. This will minimise potential traffic

disruption to one period rather than separate work being carried out by the maintaining authority after completion of the widening.

28. Integral Structures The resistance of all bridges to the aggressive affects of the environment and road salt is highly dependent on the structural form, standard of design and construction and the level of maintenance. The publication of the DOT's study of 200 Trunk Road Bridges (Ref. 7) highlighted the extent of the problem. The report identified that the use of joints at intermediate and end supports can have a detrimental effect which may eventually lead to a maintenance expenditure. This could eventually exceed the original cost of the bridge. Current work on the M6 and M5 Midland links in Birmingham is just one example of the problem facing bridge engineers involved in maintenance. The DOT are currently considering the publication of a new standard on "Design for Durability", which will require the adoption of integral bridges, ie, structures without joints, for all new structures. This practice has been adopted in several states in North America for many years and evidence shows that any small additional construction costs are quickly recouped. Over 95% of the proposed underbridges have lengths less than 80m, the critical length beyond which joints may be introduced. These bridges will need to be designed with decks tied to the abutments. This approach should present no real difficulty for the designer, but the need to keep side road traffic operating will present several challenges.

Landmark Structures

29. The Good Roads Guide, (Volume 10 of the Design Manual for Roads and Brides (Ref. 8)), identifies the benefits of landmark structures. These structures are located over or adjacent to the motorway and provide some visual relief for car users, breaking up long stretches of motorway with few features. Natural geography and road alignment may offer sufficient variation, but it is recommended that selected sites are chosen and opportunities are exploited where appropriate. The greater size and scale of structures needed for widened motorways, particularly the overbridges, means that solutions such as cable stayed bridges can be considered as economical options. Given the standardisation of the overbridges, it is even more important that potential landmark sites are identified and feasible solutions investigated. The additional effort is usually well rewarded on completion.

Conclusions

30. Motorway widening projects present many new challenges for the Department of Transport, consultant and contractor. The much larger scale of widening existing dual three lane motorways to dual four standard by parallel widening techniques means that a number of new solutions must be developed. The need to consider future modification of the motorway within the lifetime of the

structures should be taken into account so that total reconstructions are avoided if possible. New requirements which aim to improve bridge durability will also demand new solutions. When all these factors are considered together with the high profile of bridges on these heavily used motorways quality solutions are demanded and the skill of the bridge designer will continue to be tested.

Acknowledgements

31. The Authors wish to thank Mr A Whitfield, Road Programme Director of the Department of Transport for his permission to publish this paper.

REFERENCES

1. JEFFERS E. and HEALEY T.N. Motorway Widening Factors to be considered in Bridge Replacement. Construction Marketing Seminars on Bridge Replacement, May/June 1991.

2. ILES D.C. Replacement Steel Bridges for Motorway Widening. The Steel Construction Institute, 1992.

3. WILLIAMS SIR OWEN and WILLIAMS O.T. Luton - Dunchurch: Design and Execution Proceedings. Paper No. 6435 (Institution of Civil Engineers, London) May 1960.

4. DEPARTMENT OF TRANSPORT. Technical Memorandum BD 34/90 Technical Requirements for the Assessment and Strengthening Programme for Highway Structures Stage 1 - Older Short Span Bridges and Retaining Structures.
Design Manual for Roads and Bridges Volume 3. Department of Transport, London September 1990.

5. MANNING D.G. Effects of Traffic-Induced Vibrations on Billage Deck Repairs. NCHRP Synthesis of Highway Practice 86 Transportation Research Board, National Research Council, Washington D.C. December 1981.

6. BRITISH RAILWAYS BOARD. British Rail Construction Conditions. Civil Engineering Department Handbook No. 36. May 1990

7. E.J. WALLBANK. The Performance of Concrete in Bridges. A survey of 200 Highway Bridges. HMSO, London. April 1989.

8. DEPARTMENT OF TRANSPORT : Technical Memorandum HA 58/92 The Good Roads Guide New Roads The Road Corridor. Design Manual for Roads and Bridges Volume 10. Department of Transport, London. December 1992.

Demolition and removal of existing prestressed concrete structures

P. LINDSELL, Principal, Peter Lindsell & Associates

SYNOPSIS. The demolition of concrete bridges is normally performed quite independently from the design and construction phases. However, the modern bridge containing reinforced and prestressed concrete may need a detailed assessment and reference to the construction sequence in order to determine a safe demolition procedure. It is suggested that prestressed concrete structures be classified into three groups, reflecting the complexity of the structure and the information required for its safe demolition. For the more complicated structures, it is recommended that demolition drawings and a detailed demolition sequence should be prepared and closely supervised by an experienced demolition engineer.

INTRODUCTION

1. The introduction of precast concrete and, in particular, prestressed concrete has led to major changes in design and construction techniques. Many types of prestressing systems were developed and built into a wide variety of concrete bridges in the UK during the period from 1950-1970. The variety and type of prestressing system kept pace with demands for more applications and larger prestressing forces. Plain high tensile wire systems were superseded by low relaxation high quality strands and high tensile bars.

2. This rapid development has led to a new range of problems for bridge engineers and demolition contractors alike. Not only is a concrete structure likely to contain a hidden form of prestressed concrete, but the type of system, tendon details and anchorage devices may be unknown to young engineers.

3. Many of these early prestressed concrete structures are now approaching the end of their useful life. The effects of corrosion, chemical conversion, structural failure and accidents have already prompted premature demolition and strengthening of old prestressed bridges. Increases in traffic loading and motorway widening schemes have also created the need to develop methods for structural alterations to existing bridges. This paper reviews the major problems encountered in the UK over the last 20 years and describes the

Bridge modification. Thomas Telford, London, 1995

methods that have been used to encourage safe site practice and to exercise good control over the demolition process.

ASSESSMENT OF EXISTING STRUCTURES

4. A structural assessment of an existing bridge is a fundamental requirement in the planning of structural alterations or demolition of any form of structure. A basic assessment of prestressed concrete structures destined for total demolition should be aimed at establishing the prestressing systems, and the location of tendons at critical sections and proposed points of severance.

5. The condition of prestressing steel and integrity of grout in post-tensioned ducts should also be determined prior to preparing a detailed method statement for demolition. In the fully grouted condition, a tendon can be cut without serious risk of injury to an operative or the ejection of an end anchorage. However, under no circumstances should the tendon wedges or end anchorages be disturbed during a survey or before the full prestress forces have been released.

6. A rigorous assessment of complex bridge decks requiring demolition or any post-tensioned element undergoing structural alterations should be made. The presence of untensioned reinforcement is a vital factor in the behaviour of a structure undergoing demolition. The absence of longitudinal untensioned steel in precast segmental beams is particularly important as the release of prestress may cause an instantaneous brittle collapse.

7. The original design prestress and the location of tendons should be established where possible from existing construction drawings. The actual location of the tendons at critical sections may need to be established in case duct flotation has occurred during construction (ref. 1). The remaining level of prestress in a structure can be estimated in accordance with normal codes of practice, using any additional construction information arising from a condition survey. If the actual loss of prestress is in doubt or essential to structural alterations, then principal concrete stresses may be estimated using instrumented concrete coring techniques (ref. 2-3). Standard concrete cores of just 75mm or 150mm diameter are needed for this method and concrete stresses can be estimated within an accuracy of ± 1 N/mm^2.

8. The assessment of a prestressed bridge undergoing demolition, repair or alteration should include the temporary supports or bracing that may be required to carry out the works. The design of such temporary works should make special provision for the likely impact loads or resonance effects that may develop during demolition or selective cutting operations.

9. Part of the assessment process at the planning stage should consider the safety aspects of the proposed works. In particular, a careful assessment should be made of the possible consequences of a sudden loss of prestress if grouting of tendon ducts is incomplete. This check is very

important if little or no untensioned steel crosses the joints between precast units. The sudden collapse of several structures has already occurred in service where such conditions prevailed.

STRUCTURAL ALTERATIONS

10. Following the assessment of an existing bridge, it may be necessary to strengthen particular prestressed sections. A common technique to provide additional strength in a beam section is to insert short lengths of vertical and horizontal prestressing to enhance the shear and bending capacity.

11. It is possible to use either bonded prestressing systems or unbonded tendon designs for this type of repair. New additional web concrete may be added to the sides of an existing web to protect the tendons and enhance the stiffness of a beam. High tensile steel bars with threaded ends are particularly useful for short lengths since the losses of prestress at the end anchorages are very small. Alternative tension materials may also be used for external strengthening of existing sections, offering a durable and easily inspected system that may be replaced or re-stressed in the future.

12. Any alterations to an existing structure may require new openings to provide access points for services and inspection. Access holes are commonly produced by stitch drilling using overlapping cores. However, large holes up to 700mm in diameter may be removed in one section with modern core drills. Alternatively, combinations of cutting methods using core drills, diamond blades or diamond coated wires can produce holes of any size. It is important that the position and size of any hole created in the web of a prestressed beam is carefully checked in advance. Local stress concentrations will inevitably occur, even around the perimeter of a circular hole, and a limited amount of microcracking is to be expected.

13. Alterations and repairs to damaged prestressed structures often require the removal of areas of concrete adjacent to the prestressing tendons. The techniques for cutting the concrete and exposing tendons are highly specialised and expensive. Diamond core drilling and sawing will cut the prestressing steel. Water jetting with a grit additive will also cut the prestressing tendons. To avoid damage to the steel, water jetting without an abrasive additive must be specified.

14. Partial demolition of a prestressed concrete structure can also be conveniently carried out by mini-blasting combined with plain water jetting. The mini-blasting technique employs small explosive charges in a series of closely spaced small diameter holes. The technique has been used on prestressed concrete bridges, to remove chloride affected concrete from the parapet beams. Large scale removal of concrete can be achieved with little damage, vibration or noise. Two further advantages of the technique are that bridges can remain in use throughout the process and the existing steel reinforcement may be re-used.

PART 1: CONSTRUCTION

15. All repairs to damaged prestressed sections require very careful preparation of the concrete surfaces. Water jetting or sand blasting are generally suitable for this purpose. Concrete repairs should be carried out using cementitious materials with "non-shrink" properties to minimise tensile stresses in the repair material. Pourable concrete mixes are available which do not need any mechanical vibration for compaction. Sprayed concrete may be an appropriate repair material, but mechanical anchorages and a fine reinforcing mesh may be needed to avoid subsequent spalling of the repair layer.

PLANNING A DEMOLITION SEQUENCE

16. Long before a demolition contractor is appointed, the planning of a demolition sequence should commence. The main decision to be taken for many bridge structures is whether large sections can be removed with the prestress intact and the tendons released in relative safety on the ground. Alternatively, it may be necessary to prop the section and all inter-connected members to maintain stability while the prestress is released in situ. It is good practice to outline several procedures based upon either approach so that demolition contractors may evaluate the most economic option to suit their particular skills and equipment.

17. The demolition procedures outlined by the project engineer at tender stage should indicate the likely techniques that are acceptable, bearing in mind local site constraints and safe working practices. To achieve satisfactory safety standards, it is suggested that clients should issue demolition drawings and a specification to a select list of experienced contractors in the same manner as a construction tender. It follows that the removal of a complex prestressed bridge should also be fully supervised by experienced resident engineers with specialist advice being provided by qualified demolition engineers.

18. To assist the potential demolition contractors, the essential construction details and prestressing records should be summarised in the form of a demolition drawing. The location, overall dimensions and inter-connection of all structural members should be detailed where possible. It is particularly important that a complete schedule of prestressed elements with details of the prestressing tendons, initial forces and location of end anchorages is included on the drawings.

19. The specification for a prestressed concrete demolition project should clearly define the scope of the work to be undertaken. On some complex structures, it is likely that some traditional demolition techniques may not be appropriate and these should be specifically excluded. In addition, the resident engineer may require special access to parts of the structure and the provision of equipment to monitor the demolition process. The contractor should be given the opportunity to assess such items for the cost estimates and

their effect upon the programme of work.

20. A preliminary method statement at the time of tendering, followed by a detailed statement after full site investigations have been completed are essential steps in the planning process. The detailed statement should provide all appropriate calculations and checks made on the stability of the structure at each stage of demolition. The design of any temporary works and special end protection to post-tensioned members should be included, together with confirmation of the proposed demolition techniques and a programme for the demolition sequence.

21. The demolition of an old post-tensioned bridge may be particularly hazardous where poorly grouted tendons have been subjected to long term corrosion. Therefore, part of the planning for a demolition sequence should include a preliminary survey by the project engineer or the appointed contractor, in advance of a final agreed method statement.

22. A fundamental part of the final planning operation should include a specification by the resident engineer of all critical operations that require prior approval before demolition or cutting operations can commence. Such operations may include construction of temporary works, safety precautions, de-stressing procedures and handling of prestressed elements. It is always imperative that the exposure of prestressing tendons and cutting of tendons is only conducted under close supervision by an experienced bridge engineer.

CLASSIFICATION OF PRESTRESSED STRUCTURES

23. The preparation of demolition procedures may be simplified by introducing a form of classification for structures in order to restrict the amount of information stored and the effort spent on preparing superfluous specifications and drawings.

Simple prestressed beams

24. Pretensioned concrete beams may be removed complete, cut up into smaller lengths in situ or broken up with a ball and chain if the situation allows it. If cutting is necessary, then the ideal information for ensuring a safe procedure would be the member weight, the superimposed loadings, the prestress force, the number and distribution of the prestressing strands. The falsework, cutting positions and the cutting sequence could then be specified with confidence.

25. Post-tensioned beams may be treated in a similar manner, but additional information is desirable. The cable profiles, the type of grouting, size of the cable ducts, the type of end anchorages and the concrete cover to the anchorages are needed to prepare a safe demolition procedure. As an absolute minimum precaution, it is essential to know whether the tendons are bonded or unbonded and the order of magnitude of the prestress forces.

PART 1: CONSTRUCTION

Composite and continuous prestressed beams

26. Both pretensioned and post-tensioned beams may be designed to act compositely with a concrete top slab and may be continuous over several column supports. Post-tensioned structures are often stressed in stages, to act as a simple or continuous beam, stressed again after an additional composite slab has been added and stressed still further if superimposed loadings require it.

27. It is essential that this type of design and construction information is passed to the demolition contractor in order that the temporary works and cable cutting procedures can be properly designed. An accidental collapse in one part of a semi-demolished structure could lead to a progressive failure, if the inter-dependence between spans and component parts is not fully understood. It would be necessary to prepare sufficient drawings to show the full extent of the temporary supports, a detailed specification of the removal of dead load and the exact sequence for releasing the prestress forces.

Complex prestressed structures

28. This category would include post-tensioned beams that contain many cables which have been stressed progressively during the construction process. Other special structures that have been stressed in two or three directions would all be considered in this group.

29. The final order of load removal and cable cutting may only be determined after checks have been made on the stability and local stresses at intermediate stages in the demolition. Where a large group of cables are being released in a section there is likely to be a gradual transfer of load into the remaining cables and a significant increase in the steel strains can arise due to the elastic recovery of the section. This type of secondary effect can be easily allowed for in the demolition procedure providing the essential design details are available, but due allowance must be made for the possibility of ungrouted, partially corroded tendons which may fail without warning.

30. Thus, in these circumstances the demolition contractor should not be expected to proceed without any help from the client. He should receive detailed specifications of the techniques and demolition sequence with full working drawings showing essential safety measures and temporary works. He should also be supervised closely on site by properly trained engineers who understand both the design and the demolition aspects of the work.

DEMOLITION TECHNIQUES

31. Suitable techniques for prestressed concrete demolition depend primarily upon whether the entire bridge is to be removed or whether selective demolition is required. The location of a structure will limit the methods that can be employed.

32. The demolition of a bridge adjacent to existing property requires very strict control over dust, noise, fumes and vibration. Such restrictions are unlikely to be satisfied by impact breaking and traditional demolition by balling. Therefore, demolition techniques are likely to be more costly and time consuming. In contrast, the removal of an existing motorway overbridge demands the fastest possible speed and night-time working with no damage to the adjacent carriageways or danger to nearby traffic. These conditions lead to the maximum use of heavy impact breakers and hydraulic shears and occasionally to the application of controlled explosives.

33. All structures and heavy bridge decks can be most effectively brought to the ground with explosives. However, it is important to recognise that post-tensioned elements can be particularly dangerous to handle if sections are damaged sufficiently to dislodge prestressed anchorages.

34. Precise cutting techniques are likely to be more appropriate for selective demolition, structural alterations and repairs. Diamond sawing, diamond wire cutting and water jetting with a grit additive can provide comparable cutting precision. The concrete and prestressing tendons will be severed in one operation and it is important to realise that a section may fail suddenly while the remaining tendons are being cut. An operator may not be able to escape in time as collapse is possible in a few seconds.

Structural behaviour during demolition

35. Any workmen involved in demolition operations of this nature should be carefully positioned out of reach of a possible collapse and temporary props introduced to catch any falling sections. Prestressed beams are often relatively slender and special attention should be paid to lifting techniques and lateral bracing.

36. The injection of a cement grout into tendon ducts ensures an efficient bond between the tensioned steel and the concrete members. In the fully grouted condition, a tendon can be cut without serious risk of injury to an operator or the ejection of an end anchorage. However, it may be necessary to conduct "debonding trials" to assess the likely behaviour of a segmental or continuous prestressed structure. A tendon is likely to slip on each side of a cut position, causing longitudinal cracks along the line of the ducts until it re-anchors by bond action. The results from debonding trials on a range of post-tensioned beams have shown that the concrete strain changes detected on the concrete surface along the line of the tendon ducts provide an indication of the extent of debonding from the cut position as individual strands are severed (ref. 2). The length of debonding and the corresponding loss of prestress in the section is primarily dependent upon the tendon grouting, shear links, adjacent tendons and the concrete cross-section.

37. Simple pretensioned beams and slabs should present no special problems for demolition or cutting operations as the

prestressing tendons should be fully bonded to the section. Total demolition can be carried out in situ by balling, impact hammer, hydraulic shears or explosives. When individual tendons have to be cut or burnt by hand, the operator must be positioned well to one side to avoid the risk due to breaking wires or strands.

38. Many forms of post-tensioned structures exist and the likely behaviour of a structure undergoing partial or total demolition should be carefully assessed before demolition methods are selected. Structures which are incrementally stressed during construction often need to be dismantled in a progressive manner. The removal of dead load and release of prestressing tendons should be a carefully phased operation and may require specialist site monitoring to ensure that adequate factors of safety are being maintained (ref. 3).

CONCLUSIONS

39. There is still little experience available on the demolition of some recent forms of prestressed concrete construction. However, if a careful assessment of a bridge structure is made prior to demolition, it is possible to release the strain energy in the tendons by a wide variety of conventional techniques.

40. The most important requirement is maximum safety for the general public and the demolition team. Safety can only be ensured if the client, engineer and contractor fully recognise the dangers and work closely together in the planning and supervision of the demolition of prestressed concrete bridges.

REFERENCES
1. LINDSELL P. Demolition of Post-tensioned Concrete. Concrete, January 1975, vol. 9, No. 1, pp.22-25.
2. BUCHNER S.H. and LINDSELL P. Testing of prestressed concrete structures during demolition. IStructE/BRE Seminar on Structural Assessment - Based on Full and Large Scale Testing. Butterworths, Watford, 1987, p.46.
3. PRICE W.I.J., LINDSELL P. and BUCHNER S.H. Monitoring of a post-tensioned bridge during demolition. IABSE Colloquium - Monitoring of Large Structures and Assessment of their Safety, IABSE Report, vol. 56, Bergamo, 1987, pp.357-365.

New and widened bridges for the M5 motorway

I. MARSHALL, Associate, Howard Humphreys and Partners

SYNOPSIS. The northernmost section of the M5 was one of Britain's earliest motorway projects. It was built as two lane dual carriageway for much of its length. This paper describes bridge replacement modification for the widening contracts between Junctions 4 and 8. Howard Humphreys made simplicity of construction of overbridges a key feature. Maximum effort at both design and construction stages was put into reducing disruption of the traffic on an already overcrowded motorway. In all, the projects involved the strengthening, widening or complete replacement of 46 bridges and the lengthening of 11 culvert structures. Construction lasted from 1985 until 1993.

INTRODUCTION

1. The M5 Birmingham to Exeter Motorway was one of Britain's earliest motorways. It was built in several distinct phases; the first of which was the section from Birmingham to the eastern terminal of the Ross M50 Spur Motorway near Strensham in Worcestershire. The section from south west Birmingham to Strensham, unlike the sections either side of it, was built as two lane carriageway with, by today's standards, narrow hardshoulders. The pavement strength of the shoulders was substantially lower than that of the main carriageway.

2. The motorway between Junction 4, Lydiate Ash, and Junction 8, Strensham, shown in Figure 1, was constructed in a 2¼ year period between 1960 and 1962 at a cost of £250,000 per mile. The two junctions are 44 km apart and this section of motorway runs roughly parallel to the A38 Trunk Road for most of its length.

3. By the mid 1970's peak traffic densities had increased to full capacity. The M5 is a popular holiday route from the North and Midlands to the south west of England and this section of M5 suffered serious bank-holiday and summer weekend traffic delays. The A38 did not offer a suitable relief route because it passed through Worcester and Bromsgrove.

PART 1: CONSTRUCTION

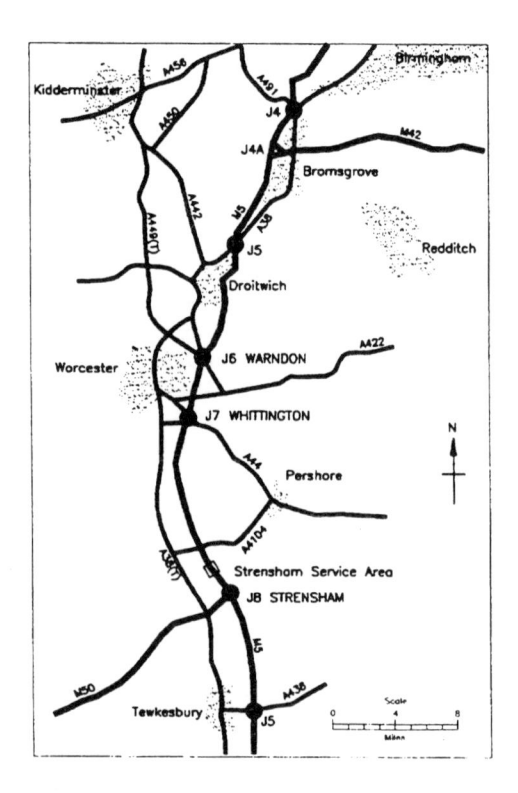

Fig. 1 : M5 Motorway in West Midlands

4. In 1980 design for increasing capacity by widening to three lanes was started by Hereford and Worcester County Council, the DTp's Agent Authority. The first scheme undertaken, from Junction 4 northwards to Dayhouse Bank involved no bridgeworks. New retaining walls were constructed to minimise the amount of land acquisition. All subsequent widening contracts, four in all, involved major bridgeworks and this aspect, together with the need for carriageway reconstruction, came to dominate the planning, design and construction processes.

5. The four widening projects which are the subject of this paper are shown below in Table 1.

Table 1. Bridges on M5 Widening Contracts

M5 Widening Project	Value Tender £m	No of Bridges	Bridgeworks Designers
1. Warndon to Rashwood	17.14	12 U/B	HWCC and HHP
2. Rashwood to Catshill	17.0	6 O/B 9 U/B	HWCC and HHP
3. Catshill to Lydiate Ash	20.4	5 O/B 2 U/B	HHP*
4. Warndon to Strensham	59.4	19 O/B 5 U/B	HHP

* The M42 terminal interchange with M5, Junction 4A, was designed by Ove Arup and Partners and was incorporated into Contract 3.

6. Contracts 1 and 2 were executed under ICE Fifth Edition Conditions of Contract with Hereford and Worcester County Council and Howard Humphreys and Partners named as Joint Engineers. Because of the inclusion of Junction 4A into Contract 3, its designers, Ove Arup and Partners, also joined the named engineers making it a three Engineer contract. Each Engineer spoke on those parts of the works for which he was responsible but when an issue affecting the whole contract, such as a Extension of Time arose, HWCC acted as spokesperson for all three irrespective of the origin of the need for a statement. The system for each contract worked very effectively.

7. Design and contract supervision of the fourth and largest contract was awarded to HHP after a DTp fee competition. The project was opened as full three lane motorway on 30 July, 1993.

CONSTRAINTS ON DESIGN

8. As well as the normal array of standard design requirements the design of bridges for the M5 widening projects had to take into consideration existing traffic, geometric and materials conditions in earthworks, pavement and structures.

Traffic

9. The view taken originally by Howard Humphreys and Hereford and Worcester was that successful management of the motorway traffic would be the key issue for M5 widening. Keeping traffic flowing safely and as quickly as possible with minimum disruption or distraction was seen as essential. As

the schemes progressed the importance of the original view became clearer and more and more emphasis was placed on it in successive schemes.

10. To ensure that the goal of minimum disruption was met any design proposal which might have resulted in parts of live carriageway being closed off were challenged and tested before incorporation. This led to integration of the construction programmes for carriageway and structures very early in the design process. Preparation of technical approval applications became a very interactive process between highway and bridge design teams.

11. For example, it was found that in the majority of cases the cost savings which would have been made by use of central reserve piers were outweighed heavily by the actual costs of installing and maintaining lane closures plus the delay costs predicted by QUADRO analyses.

12. As the designs moved from project to project the 'acceptable' restrictions were modified progressively as lessons were learned. The first contract was designed with contraflow working and traffic reduced to single lane at certain times. Subsequent designs were modified, first to maintenance of two lanes (as existing) in each direction at peak seasons and then to maintaining them at all times. The fourth, and final contract differed radically by adoption of the 'parallel' widening format. As well as allowing very rapid construction this gave uninterrupted flow on the existing carriageways for the maximum time, kept the existing number of lanes open and, very importantly, maintained continuous safety fence or other barrier between opposing flows at all times.

13. In all the forms of widening, demolition of existing bridges and construction of the new ones sometimes required complete closure of the motorway. Since this was possible only during limited numbers of night closures this also became a major constraint on bridge design proposals.

Geometry

14. Omission of hard shoulders at bridges, a feature of some very early motorway widening projects, was never considered as an acceptable solution. Once the decision to have full 3 lane carriageway and hard shoulders throughout the widened length of motorway had been taken, replacement of all existing overbridges was inevitable. None of the bridges had sufficient clearance for the new road layouts and in some cases overspanning was needed to accommodate extended slip roads at upgraded interchanges.

15. Providing that there are no major obstructions within a few metres of an existing deck edge it should, in theory, be possible to widen most types of underbridge. However, on the first three M5 projects geometric or structural problems meant

that some underbridges were replaced rather than extended. The alterations to the geometry of the last contract, Junctions 6 to 8, were considerable and every bridge was replaced.

Testing Existing Materials

16. Detailed materials appraisal was carried out on any of the existing bridges being considered for re-use. The process started with close inspections of the structures followed by a regime of materials investigation and testing.

17. The investigations included testing for chloride contamination, electro-potential half-cell activity, resistivity and carbonation depths. With the exception of very small localised areas of chloride contamination, all bridges were assessed as having low risk of future reinforcement corrosion.

18. A number of cases of alkali-silica reaction had been found previously in the West Midlands area. Some of these were in highway structures. As a result it was decided to carry out ASR testing on the M5 bridges. It was found that a number of underbridge substructures were of 'medium' susceptibility and in some cases cracking had already occurred.

Strength Assessments

19. Strength assessment of any of the structures which might be retained in the new schemes was carried out prior to technical approval application. As the structures were to be part of a 'new' motorway the assessments were for compliance with current new design standards and not to the bridge assessment standard, BD 21/84.

NEW OVERBRIDGES

20. Although they are the most conspicuous structures to motorway users the replacement overbridges have, in general, been easier to deal with than underbridges. Two main types of overbridge were adopted for three of the contracts (one contract had underbridges only)
    a) single span
    b) two span asymmetrical

Adoption of parallel widening for the final project led to extensive use of the asymmetrical two span layouts. Elevations of the two main types are shown in Figure 2.

PART 1: CONSTRUCTION

Single span

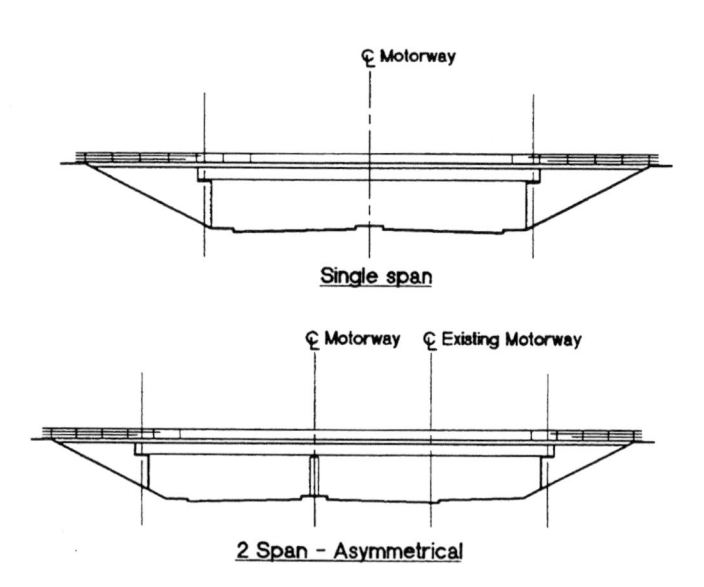

2 Span - Asymmetrical

Fig. 2   M5 Overbridge Elevations

21. At all the overbridge sites on M5 it has been possible to locate the new bridge alongside its predecessor thus allowing the road over to remain open throughout construction. Completion of tie-ins and transfer of traffic to the new bridges has usually been achieved with minimal disruption although at one site diversion of a major water main required an 8 weeks closure of the side road.

Sub-structures

22. Earlier, reference was made to possible traffic delay costs inherent in constructing central reserve piers. This applies only to the existing reserve. Two of the bridges between Junctions 4 and 5 fell into this category but, as the construction of new link roads for the M42 to M5 allowed brief access to the reserve, they could be built without incurring the penalty of further traffic delays. Construction of intermediate supports for parallel widening took place off-line. Only in the final traffic layout did they appear in the central reserve.

23. Clearance to excavations was considered carefully through the design stages. Desirable basic practice for pier construction is summarised by Figure 3. In reality, there has not always been sufficient space to achieve these ideals. Sheet piling has sometimes had to be used to reduce space required for foundation excavations. In more extreme cases the haul road has had to be omitted and site vehicle access past the bridge has been sought elsewhere.

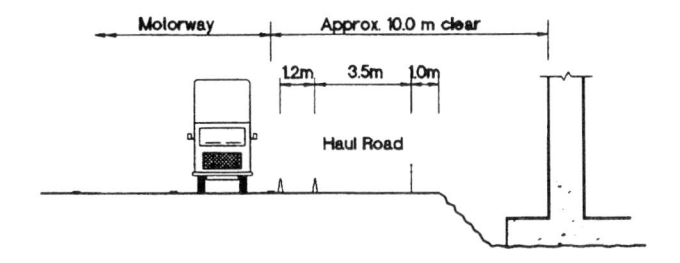

Fig. 3 : Clearances to Substructures

24. In general, clearances for safety requirements have become more onerous with each new project. A major step forward has been the widespread introduction of vertical concrete barriers which give enhanced protection to the workforce. Their use, however, is restricted to sites with mandatory 50 mph speed restrictions and, because they require 600mm set back from the traffic, offer limited savings on space.

25. Very simple substructure sections have been adopted for all four M5 widening projects. Sections and profiles which would have required significant strutting or guying were avoided to obviate obstruction to the limited access available. Single or double leaf piers have been used throughout the four contracts.

26. Abutments generally consist of full height reinforced concrete cantilever walls with wingwalls in line with the structure. This minimised the clearance between new and old structures and the necessary deviation of the side roads. Coincidentally, this put some of the side roads back on their original alignments, lost when the M5 was first built. On the parallel widened section earthworks were extended around the new abutments to reduce the area of exposed concrete and give the impression of a bank-seat type structure. This gives the further advantage of being removable to facilitate future alterations to carriageway layouts should the need arise.

27. With a few exceptions, mainly at the extreme northern and southern ends of the fourth contract, ground conditions were good and spread footings could be used directly on to the underlying marl. Control of water to prevent deterioration of the marl, once exposed, was essential.

Superstructures

28. The need to span the existing motorway without restrictions governed the main span of 31 metres. For the parallel widening layouts the second span needed only to span the new carriageway. However, in order to balance the spans and avoid uplift the second span was set at 22.5 metres. These figures refer to square spans; for skew crossings the actual

PART 1: CONSTRUCTION

beam lengths increase correspondingly. Bridges crossed the motorway at up to 40° skew giving rise to spans over 40 metres at some locations.

29. <u>Steel concrete composite</u>. All of the overbridge decks on the four contracts are of composite steel and concrete construction. The steel plate girders have been adopted in all but a few cases where box girder construction was more suitable. The use of insitu concrete decks requiring extensive falsework over the live motorway was ruled out at a very early stage. In appraising the relative merits of various forms of construction account was taken of the difference in cost of cranage for steel girders and the much heavier precast concrete units which would have been needed for such long spans. The increased time for rigging the larger cranes was also considered when they would have stood on the carriageway.

'BRACED PAIRS'

30. Bridges were designed with even numbers of plate girders enabling them to be assembled in pairs to aid stability during lifting and deck slab construction. Grade 50 steel was used throughout. Whilst multiple pairs were used for most bridges the four accommodation bridges between Junctions 6 and 8 were designed and built as single pairs of plate girders.

31. <u>Falsework</u>. More efficient use of motorway closure time was sometimes made by fitting parapet falsework in place to line and level on the outer pairs of girders prior to lifting. Temporary decking/screening was often fixed to bottom flanges before lifting to save on erection time.

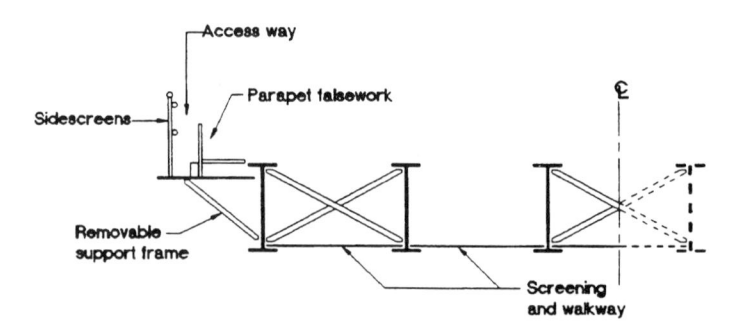

Fig. 4 'Braced pairs' of overbridge girders

32. <u>Girder Erection</u>. Figure 4 shows the very basic cross section of deck lifted into place during a typical overnight possession. The maximum number of pairs installed in a single night was three. Splice connection details for multispan structures were made as simple as possible because all had to be assembled in mid-air and frequently at night. Wheeled

scaffolding towers were usually used for access by the steel riggers. These could be removed rapidly once splice bolts had been tightened fully. Soffit screening on flanges between adjacent braced pairs was slid into place as soon as the girders were landed and levelled. Deck slabs for the final project were designed with permanent concrete plank formwork to give reduction in time spent working over the motorway by removing the need for screening. Pretreatment of the concrete planks with silane gave yet further savings in construction time.

INTERCHANGES

33. New overbridges were required at only two interchanges, Junctions 6 and 7. At these sites roundabouts over the motorway would have meant provision of four very wide structures unless the bridges were curved in plan. Plate girders formed from four straight sections, "twenty piece" fashion, were used to narrow the decks. The concrete deck slabs were then cast with curved edges to suit the interchange geometry.

REPLACEMENT UNDERBRIDGES

34. Unlike the new overbridges most of the replacement underbridges built for M5 have had to be built on-line. Fortunately, short closures of the roads crossed by the bridges designed by Howard Humphreys were permitted. Design, detailing and construction sequences of each bridge were structured around these closures and the motorway traffic requirements.

35. Because of the wide variety of obstacles crossed, spans, skews and so on the underbridges lacked the similarity of key features that the overbridges had. Existing bridges decks were reinforced, pretensioned, post-tensioned and reinforced concrete types and substructures were equally varied. Consequently, the types of replacement bridges have almost equally varied, including steel concrete composite, M beams and inverted T beams.

Construction Phasing

36. One feature common to all the on-line replacement bridges has been the complex construction phasing adopted. For example, in order to satisfy the traffic requirements whilst replacing the four of the underbridges between Junctions 5 and 6 it was necessary at times to work within 'island' sites with motorway traffic running on either side. Quite apart from safety considerations, this was less than ideal because productivity was limited by the size and type of plant which could be used in such confined areas. However, unless sufficient space exists for diversion of all motorway traffic on to one carriageway the creation of island sites is inevitable.

PART 1: CONSTRUCTION

37. Demolition of the existing bridges required programming into each of the partial construction phases. Careful assessment of the existing components still supporting traffic during intermediate stages was essential.

Substructures

38. Two basic types of substructure were adopted for bridges replaced in-situ. The first of these were of bank-seat type founded either on bored piles or mass concrete filling down to a hard marl stratum. They were built behind the existing abutments so that the new bridge overspanned the old as shown in Figure 5.

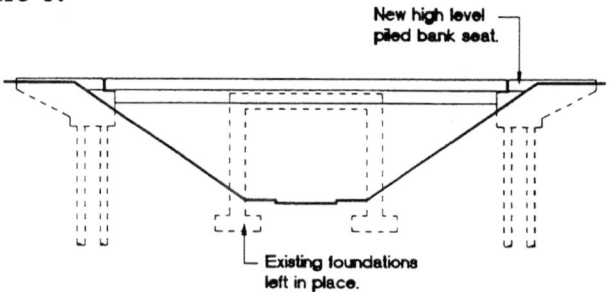

Fig. 5 On-line replacement of underbridge

Adoption of this design layout gave minimum construction time in the motorway carriageway by restricting excavation depths as well as limiting the side road closure times.

39. The second type of substructure was used at the redesigned M5/M50 interchange. This site was much less constricted than the previous ones. Realigned motorway links gave opportunity to divert a whole carriageway and allowed access to half the bridge at a time. As the other motorway carriageway remained in place it was still essential to limit excavation depths next to it. This was achieved by adopting a design suitable for top down construction with full height contiguous bored piled wall abutments and a deck built at ground level.

Superstructures

40. Because, at various times, motorway traffic was required to run adjacent to the edge of incomplete decks provision for installation of temporary P1 parapets to protect vehicles was included in the Contracts. Sometimes this was achieved by using specially made steel parapets and at other sites the permanent aluminium parapet components were used temporarily.

41. DTp Internal Memorandum IM5, requiring periods of restricted or no loading on the deck to control traffic induced movement in the immature concrete in the decks, caused considerable problems on the earlier contracts. Scope for

diverting traffic away from the joint areas was small and so traffic restrictions became unavoidable. Great efforts were made to avoid similar problems affecting the later contracts.

WIDENING AND STRENGTHENING

42. Where there was opportunity to widen structures, and this was confined to the earlier contracts, a simple principle was adopted. That was, to ensure that new construction type was as close as possible to the original in order to maintain the same structural behaviour. Thus, cantilever abutments were extended with new cantilever sections, propped/pinned abutments with propped/pinned extensions and slab decks with slab extensions.

43. Ensuring that the extensions did not shed unacceptable load back into the original structures was problematical especially concerning differential movements between substructures. To overcome this a two stage construction approach was used at five of the bridges in the first two contracts.

44. First the abutment stem was constructed with a horizontal cantilever flap at its rear face for full height. Backfilling and compaction up to deck level was carried out. Next the deck extension was built, leaving an open joint strip between new and existing parts. By using this sequence the extension foundations, which were marls and sands, were subjected to up to 90% of their final permanent load effects prior to the final operation which was to make permanent structural connection to the existing bridge by concreting the joint strips in abutments and decks. Figure 6 shows a typical accommodation underpass joint strip.

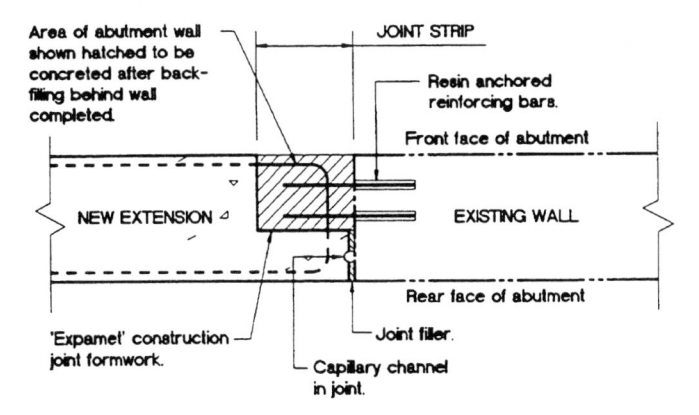

Fig. 6 Plan on Abutment Joint

45. Strengthening measures were eventually used on only farm accommodation underpasses. Their assessed deficiencies included inadequate flexural strength in abutment stems,

PART 1: CONSTRUCTION

inadequate transverse hogging moment capacity in decks and inadequate live load shear capacity in decks

46. The abutments are propped apart and pinned top and bottom and have been strengthened by external bonded steel plate reinforcement on their front faces. The design method used for the plating used the principles set out in TRRL research papers published in 1984 but with longitudinal shear stress on the adhesive/concrete interface limited to 0.41 N/mm². Additional shear resistance was provided at the ends of each plate by resin bolting the plates to the concrete.

47. Both of the deck deficiencies were overcome by overslabbing the decks with reinforced cast in-situ concrete connected to the existing deck top by means of vertical rebar shear connectors. The greater construction depth enhanced dispersal of wheel loads and increased effective section depths giving increased live load capacity. The overslabbing also allowed inclusion of a new top mat of reinforcement which produced greatly improved flexural hogging capacity in the slab.

DEMOLITION

48. By far the greater part of demolition carried out has been removal of existing overbridges. The planning of this work was given very close attention throughout design and construction planning stages. Great effort was put into obtaining all available information about the bridges and where details were not clear, additional investigations were undertaken. The overbridges removed all had concrete superstructures and varied in configuration and size, from single span post tensioned to four span in situ slabs, with the largest deck being some 1000 tonnes of precast beam and insitu concrete infill.

49. Almost every bridge was demolished by means of controlled explosives which were used both to bring down and to fragment the structures. The method was used for all of the widening contracts by the same specialist subcontractor throughout. Following discussions with local Environmental Health Officers, Police and other interested parties detailed method statements were produced for control of events preceding each blast. Extensive public liaison was necessary to allay concern and warn local residents of the procedures formalised to ensure the safety of all concerned.

50. Although carried out late at night the 'bridge blasts' soon became popular spectator events. Such was the demand that eventually provision was required for additional marshals and a safe viewing area for up to 200 people.

51. Demolition was carried out during overnight motorway closures. On the early schemes these were limited to 11.00pm Saturday until 7.00am Sunday but were later extended to 9.00pm - 9.00am without serious effects on traffic flow off the motorway. Careful planning and close co-ordination with all

concerned enabled the work to be completed successfully with few complaints and no significant damage to adjacent property.

52. Stitch drilling and sawing decks into pieces small enough to be lifted out was also used successfully on occasions where explosive demolition was impractical usually because of very close properties. Most drilling and sawing had to be carried out at night with lane closures to deal with the very large amount of dust and lubricating water pouring on to the motorway. The beams into which the decks were cut were the heaviest components lifted on any of the projects and careful planning for space required for cranes was essential.

CONCLUSION

53. After ten years of continuous involvement with bridgeworks for motorway widening it is possible to draw some conclusions from lessons learned. Most of them stem from the fact that management of the existing traffic dominates all other aspects. This may not always be easy for bridge engineers to accept but they ignore it at their peril.

54. Awareness of the very limited space and time available for construction is essential for success. Complex details stand limited chance of successful assembly if they have to be built within a busy night-time closure. The success of using pre-assembled pairs of simple plate girders for rapid, reliable overbridge construction has now been proved beyond doubt.

55. Further refinement of construction of underbridges for reduction of traffic disruption promises to be the most rewarding area for research and innovation.

56. Apart from learning these lessons from our ten years involvement we have also gained immense satisfaction from contributing to changing the highly overcrowded and unpleasant peak time M5 journey between Junctions 4 and 8 into a pleasanter and, hopefully, safer experience.

The Kent approaches to the Dartford crossing

M. A. CHAMPION, Associate, Mott MacDonald

SYNOPSIS. This paper describes the reconstruction of bridges on the A282 Dartford Tunnel Approach Road in Kent between 1989 and 1991. It outlines the criteria considered on highway widening projects and describes the different construction methods used. The experience gained on this very heavily trafficked project is relevant to many such future schemes.

HISTORY OF THE SCHEME

1. The Dartford tunnel was opened as a single bore 2-lane crossing of the Thames in 1963. At that time the Kent approaches were of dual 2-lane construction. A rapid increase in traffic led to a demand for a second crossing which was opened as a single bore 2-lane tunnel in 1980. This effectively increased the crossing capacity to that of the approach roads. However the presence of the Princes Road interchange led to congestion as a result of local traffic entering and leaving the main approach roads. Widening of the approaches to dual 3-lane (with the exception of the length through the interchange) was completed in 1987 to coincide with the completion of the M25 motorway. No bridge replacements were carried out but retaining walls were introduced to replace cutting slopes and thereby increase available carriageway width. Although the tunnel and its immediate approaches were not designated as motorway, they formed the eastern part of the M25 orbital route.

2. Following the completion of the M25, there was another rapid rise in traffic which led to queues at the toll-booths for the tunnels. There was public demand for an improvement of the Crossing to 8 lanes and the Government decided to involve the private sector. This led to a design and construct competition in 1986 which was won by Trafalgar House on the basis of a 450m span cable-stayed bridge with a mid-span clearance of 54m located just downstream of the two existing tunnels. This competition included the river and flood plain crossings but not the approach roads.

3. At the same time, the South East Regional Office of the Department of Transport appointed Mott MacDonald as the Engineer for the widening of the approach roads so that these would correspond to the motorway and provide dual 4-lane capacity with slip roads in addition wherever necessary. This widening applied between the A2 interchange (J2) in Kent and the A13 interchange (J31) in Essex.

4. The contract for the Kent side was awarded to Balfour Beatty in May 1989, and that for the Essex side to Cementation in July 1990. Both were to finish simultaneously with the opening of the Queen Elizabeth II Bridge in November 1991.

DESCRIPTION OF THE PROJECT
5. The Kent section, which is the subject of this paper, is 4.5km long and contains three interchanges. Four overbridges were replaced and another had its deck relocated on new supports some 45 metres northwards from its original position. Of the four new bridges, two entailed constructing the decks in temporary positions offset from the existing bridges, so that these remain in use for as long as possible, with the new decks being slid into place once the old bridges had been demolished and new supports built.

6. The A282 is a motorway in all but name, although the existing road geometry was substandard in many respects. At the most restricted point on the project, Princes Road Interchange, the dual two lane carriageway with no hard shoulders was constrained by retaining walls supporting slip roads to the interchange, which were in turn also protected by retaining walls. Elsewhere on the project, standards were more generous at dual three lane plus hard shoulder. Prices Road Interchange and the A282/A2/M25 Interchange both needed to be upgraded and widened to three lanes, with new link roads provided between them. Twenty-four motorway style overhead sign gantries were also to be provided and erected through the project. All this had to be achieved on a section of road carrying approximately 100,000 vehicles per day, of which nearly 30% were heavy goods.

THE NEED FOR TRAFFIC MANAGEMENT
7. It was clear at the design stage that a high level of traffic management would be needed, the principal objective being to maintain the existing traffic flows as much as possible whilst providing access and a safe working environment for the construction force. This was particularly important in locations in close proximity to the heavily trafficked live carriageways where the congested and restricted nature of the site made access and working space very difficult.

8. Although forming the closing section of the M25 Orbital Motorway, with the traffic flows already mentioned, the A282 carries a very high proportion of local traffic with the only restriction on vehicles being that they must be engine powered. Because of its "local" nature there is a high proportion of relatively short journeys with a greater than usual use of the

PART 1: CONSTRUCTION

junctions. The A2/M2 corridor, to and from the Channel Ports and joining the A282/M25 at the southern end of the project, contributes a very high proportion of heavy goods vehicles, of both British and Continental origin, which pass through the Tunnel en route to the Midlands and North. The Dartford Tunnel itself was run as a commercial operation with toll receipts linked to the funding of the new bridge and, consequently, any interference with traffic flows had to be minimised.

9. Consideration also needed to be given to protecting road users during any overhead works and a careful assessment of the extent of the risk to public safety dictated the measures to be implemented in each case. The installation of sign gantries and new bridge deck beams over the carriageway, involving the use of large high capacity cranes, was sufficiently hazardous to warrant closure of the relevant sections of the A282 below. In practice, although overhead installations tended to apply only to one carriageway at a time, considerations for working space, such as positioning the crane across the central reserve to be able to lift from the delivery vehicle and slew into position, frequently dictated the need to close the whole road.

10. Sliding decks from initial to final locations is not a fundamentally unstable operation but there is still an element of risk, particularly from falling tools or accidents involving hydraulic hacking equipment, that merits a road closure. For all closures of the A282, alternative diversion routes needed to be identified and set out in the contract documents. The same applied to diversion routes expected to be required when works on bridges carrying local roads restricted their use by the public.

TRAFFIC MANAGEMENT PHILOSOPHY

11. With the complexity of the project in mind, it was essential to determine the practicalities of construction whilst still maintaining a high level of traffic flow. For this purpose, and in addition to the in-house experience available, construction industry sources were consulted to help produce a possible construction programme at the design stage.

12. Results from the contract documents illustrated the feasibility of construction in conjunction with staged completions. As with all proposed programmes and temporary works included in contract documents, there was provision for a successful tenderer to produce an alternative, but such an alternative would only have been acceptable to the Engineer if it was shown that traffic flows would not be adversely affected. On schemes of any complexity, it is unrealistic to expect that tenderers, in the short period of time during tendering and even up to the start of work, will be able to grasp all the essentials of the scheme with which the designer has become familiar over a period of perhaps some years. As a significant level of reliance is likely to be placed upon a designer's concepts, it is essential that these are thoroughly and completely considered.

13.   It obviously depends on the degree of planning at design stage how diversions are dealt with in the contract documents but such decisions can have considerable financial implications for the project. It is clear however that on a major project, diversion routes cannot be left to the contractor to deal with and so, once again, the provisions in the contract must be complete in every detail.

CONCEPTUAL DESIGN OF REPLACEMENT BRIDGES
14.   Historically, there has been a complete reversal in the priorities assigned to roads and bridges in any particular project. In the early years of the railway and highway booms it was normal to provide square spans wherever possible because bridges were relatively expensive, and skew spans were difficult to design and construct. The increasing need for skew, curved or shallow bridge decks in motorway projects to suit the highway alignment led to the development of modern types, any of which may prove the most economic choice for a particular "green-field" site where there are no other constraints.

15.   Progressing to bridge replacement for the widening or upgrading of motorways has altered the balance of the criteria used in selecting bridge types. Such schemes demand minimum disruption to traffic flow. For this reason overall economy is more likely to be achieved as a result of construction speed and good traffic management rather than minimum material content within a particular bridge.

16.   The fundamental needs in designing for replacement are:

- Remember that bridges are subservient to the requirements of
  - traffic management
  - the construction programme
  - service diversions

- Bridge types should not be finalised until the above requirements have been fully defined

- Bridge types must be capable of rapid construction with minimum highway possession times.

- If bridges are required early in the contract period it may be necessary for the Employer to pre-order fabricated materials

- Demolition methods, and their consequences, must be considered in the design process and it may be necessary to specify them in the contract documents.

GENERAL DESCRIPTION OF BRIDGES TO BE REPLACED
17.   There were five overbridges on the original scheme. These have been replaced by a total of eight bridges, three of which are service/foot bridges. Travelling from north

PART 1: CONSTRUCTION

to south, the original bridges were as follows:

- **The Brent and Watling Street Bridges**
  Three span reinforced concrete portal bridges carrying local traffic and major services over the main carriageways.

- **Princes Road Roundabout Bridges**
  Three span reinforced concrete portal bridges carrying interchange traffic and major services over the main carriageways.

- **Green Street Green Road Bridge**
  Three span prestressed concrete voided slab carrying local traffic and minor services over the main carriageways. This bridge had a skew of 45°.

EFFECT OF SERVICE DIVERSIONS ON BRIDGE TYPES

18. Early discussions with all the statutory undertakers established the existing services carried on each bridge, likely future requirements and the approximate times required to carry out diversions of existing services.

19. In the case of the Brent and Watling Street it became apparent that the time necessary for the diversions was such that it could only be fitted in to the overall contract programme if work was started very early. This meant having new bridges available by month 3 of the contract. Furthermore, it was not possible to interrupt the services, which meant that a replacement bridge had to be available by the time diversion work started. The only way of achieving this was by providing separate service/foot bridges early in the contract. This in turn meant pre-ordering fabricated steelwork prior to inviting tenders for the main contract so that it could be issued to the contractor for erection during the initial weeks of the contract period.

SELECTION OF BRIDGE TYPES

20. As will be seen from the figures, it was generally necessary to replace the old 3-span bridges with new 2-span structures. In addition, the position of the central reservation had to be moved for the new carriageway layout which necessitated the provision of a temporary pier in the old reservation and a permanent pier in the new reservation. Clearly, it was also desirable to have a bridge deck which was continuous over the centre support. The need for rapid erection of the primary bridge deck members indicated the use of prefabricated units rather than anything constructed insitu on falsework. It rapidly became apparent that composite construction would be the most appropriate form to satisfy all the above requirements. Also, the service/footbridges could be of similar construction, having two beams with a concrete trough cast between them prior to erection.

NEED TO PRE-ORDER STEELWORK

21.   As mentioned above, the programme necessitated erection of some of the bridge steelwork in the initial weeks of the contract period. During discussions between the Engineer and Employer it was agreed that a preliminary contract should be awarded for the fabrication, protective treatment and delivery to store of all steelwork and bearings. This material would then be available for collection and erection by the main contractor as from the beginning of his contract.

22.   A contract for the fabrication of steelwork for four road bridges and three service/footbridges was awarded to Fairfield Mabey in July 1988. Delivery of 470 tonnes of material to a store in Surrey was completed in February 1989.

23.   During fabrication, full trial erections were carried out on all bridges. This was deemed necessary to try and eliminate as many problems as possible prior to final erection during limited highway possession periods. In the event the trial erections proved worthwhile as a means of verifying cambers and rectifying minor, but time consuming, problems due to lack of fit.

24.   This form of contractual arrangement demands careful attention from the Engineer and his inspectors to ensure that the material which will be made available to the main contractor is precisely as described in the contract documents. Any delays resulting from errors will be very expensive in financial terms due to disruption of the programme.

SIDE BY SIDE REPLACEMENT (PRINCES ROAD BRIDGES)

25.   The original Princes Road roundabout was of generous proportions and it was possible to design the new roundabout such that the bridges were positioned inside the original ones. Traffic management was therefore straightforward, allowing use of the old bridges to continue while new construction proceeded alongside. Services in the old bridges had to be diverted before they could be demolished. For this reason a new service and footbridge was constructed across the carriageways in the centre of the roundabout. Pedestrian connections were achieved via extensions to subways which existed at the four "corners" of the old roundabout.

26.   Abutments for the new bridges were constructed as contiguous bored pile walls which continued along the back of the verge line as sloping wing walls. This form of construction was to be used in many places on the contract to form retaining walls and it was economic to continue its use in these locations as well as being the least disruptive method. Unfortunately it was not possible to provide adequate resistance to overturning forces and the full height walls were supplemented by king piles placed some distance behind and joined to them by ties.

PART 1: CONSTRUCTION

27. Both the temporary and permanent foundations in the central reservations were spread footings founded on the chalk. The contractor opted to use steel support systems for the temporary piers whereas the permanent piers were of reinforced concrete.

28. Bridge decks were of conventional composite construction utilising 6 beams with channel diaphragms.

29. Following completion of surfacing, furniture and finishes to the bridges and adjacent highways, traffic was diverted on to them and the old bridges demolished. At this stage it was possible to close one main carriageway and divert traffic on to the appropriate half of the roundabout via the slip roads. Demolition of the bridges could therefore be achieved in two stages by leaving one half of the central span cantilevering from the pier. Demolition was achieved by the use of "peckers and nibblers" working within scaffold enclosures designed to trap falling debris and reduce the nuisance caused by noise and dust.

SLIDE-IN REPLACEMENT (THE BRENT AND WATLING STREET)

30. The Brent and Watling Street bridges both carried local traffic across the main carriageways. The bridges had single 10m wide carriageways flanked by 1.8m wide footpaths. It was essential for local traffic management purposes to maintain these roads throughout the contract period except for possessions of short duration. For this reason the replacement was conceived on the same basis as for a railway bridge where the permanent way has to be kept in service. It was found possible in each case to make a temporary diversion of traffic on to the new bridge if it was constructed immediately alongside the existing bridge prior to slide-in.

31. In addition, services carried by the existing bridges had to be diverted on to the new crossings. Because of the time required to do this by the Statutory Undertakers, and the need to complete the work early in the contract programme, the decision was taken to provide a separate bridge for the services which would also serve as a footbridge on one side of the new crossing. The service bridges would be erected on one side of the existing bridges prior to their demolition. Services to be accommodated were as follows:-

	The Brent	Watling Street
Water	1 x 300	1 x 300
	1 x 100	1 x 150
British Telecom	18 x 120	15 x 120
Gas	450 sleeve	Nil
LEB	4 x 100	6 x 100
CEGB	Nil	3 x 150
Sewerage	1 x 200 sleeve	1 x 400 sleeve

32.   The abutments of the existing bridge had been modified during the earlier widening scheme so that they were incorporated in to the retaining walls alongside the slip roads. The soil mass behind the retaining walls in the vicinity of the abutments had been reinforced with either Pali Radici piles or barrettes. Consequently it was a relatively straightforward operation to modify the walls so that they could be used to support the service bridge in its final position and the new road bridge in its initial position prior to slide-in.

33.   One of the first operations after award of the contract was to collect the service bridge steelwork, assemble the main beams in pairs between splice positions and cast the concrete trough. Subsequently, the pairs were erected, the splices bolted and the short sections of trough cast across the splice positions. The bridges were then ready for services to be installed and diverted.

34.   The new road bridge decks were erected alongside the existing bridges on the modified abutments and temporary piers. Once the deck had been surfaced and the parapets fixed, traffic was diverted on to it leaving the existing bridge available for demolition. After demolition had been completed, together with the permanent pier and abutment works, the new bridge was closed during a possession whilst it was slid into its permanent position.

ROTATION TO REDUCE SKEW (GREEN STREET GREEN ROAD BRIDGE)
35.   This bridge carried local traffic and was well used during rush hours. However, alternative routes were available and it was agreed that the road could be closed for several months. The bridge crossed the main carriageways at a skew of 45° and had piers and abutments square to the deck. It is a prestressed concrete voided slab bridge having three spans of 23.8, 43.9 and 23.8m. The main carriageways passed through the centre span and the sidespans were unused.

36.   After considering various options for replacement, two alternatives were developed in more detail and cost estimates were prepared. The alternatives were:

(a)   Lift the bridge deck, support on falsework under all spans. Demolish the deck, piers and abutments. Construct new bridge on old alignment.

(b)   Lift the bridge deck, slide northwards and rotate in plan to reduce skew. Lower bridge deck on to new piers and abutments. Reconstruct road approaches to suit new bridge alignment.

37.   The cost estimates indicated that scheme B was slightly cheaper but it also had the advantage that no demolition of the deck would be involved over a live motorway and that no falsework (which might be susceptible to "bridge-bashing") would be required. It was agreed to proceed with

PART 1: CONSTRUCTION

scheme B but the Employer accepted the need to have a nominated sub-contractor for the slide operation and for all temporary works to be adopted as the Engineer's design.

38.   By rotating the bridge in plan so as to reduce the skew from 45° to 30°, it was possible to increase the central square span from 31m to 38m, thereby accommodating two additional lanes. Also, by using a full height abutment it was possible to use the sidespans to accommodate slip roads. A simple rotation in plan about the centre point of the bridge was not possible because the west end would have infringed the adjacent property access boundary. A move to the north was therefore necessary to accommodate the realignment of Green Street Green Road.

39.   The Engineer worked closely with the nominated sub-contractor for the slide (Heavilifts Ltd) in developing the temporary works. A design for the slide tracks using a known source of second-hand tubular steel column supports was prepared and detailed. However, the main contractor who was responsible for constructing the slide track supports proposed an alternative reinforced concrete scheme which was adopted.

40.   The slide was accomplished using "sledges" carrying two linked hydraulic jacks mounted on a stainless steel/PTFE track. Hydraulic pull-jacks were positioned at the remote end of the slide track. The system was designed to function with a coefficient of friction of 0.10 but once the bridge was moving the coefficient was of the order of 0.03. The moving speed, including stops, was approximately 5m/hr. The slide operations were programmed to take place during three night closures of the motorway; two for the transfer and one for the rotation. Due to the good progress made, the transfer was completed on the first night and the rotation accomplished during the second closure.

41.   Following the final positioning of the deck, it was lowered on to its new permanent bearings prior to fitting expansion joints.

DEMOLITION METHODS
42.   The contractor was allowed complete freedom in choosing his demolition methods within the normal constraints of safety, Engineer's consent and environmental controls. The Engineer envisaged the possibility of wire sawing into relatively large pieces but the contractor opted to use a variety of breaking and crushing plant which produced material of relatively small size. Enclosures were used wherever possible to control the nuisance caused by noise and dust. More substantial barriers were required to prevent material falling on to live carriageways when work was in progress adjacent to them. No work was permitted over live carriageways.

CONCLUSIONS

43.   It is essential to formulate, at the design stage, the practicality of a construction programme. The programme has to take into account the construction of structures, diversion of public utilities, the number of lanes to be kept in use at any time, and this all in turn leads to the stages of traffic management and types of bridges required.

44.   Consultation with local authorities, police and public utility authorities is necessary at a very early stage in design and it is important to ensure that any relevant comment received from them is incorporated into the Contract Documents. During the project, close liaison has to continue in order that information can be fully disseminated, particularly beyond the confines of the site.

45.   The guidelines available on implementation of traffic management schemes do not take into account schemes of such a complexity as Dartford and careful consideration must be given to each project individually concerning actual implementation. This, in turn, means that both the contractor and the Engineer must have experienced personnel on site with hands-on experience capable of assessing that the traffic management proposals as well as the associated demolition and reconstruction of bridges are both efficient and safe.

46.   Clearly stated requirements in relation to maintenance of the project should also be laid down, especially with staged construction where very often it is difficult to pass on areas of the project to the Agent authorities as one would normally.

47.   Generally, the more information provided to the Contractor at tender stage then the more realistic will be the policy. This in turn will help the site staff in their management of complex projects.

48.   Finally, there are many ways of replacing bridges. The requirements for replacement are often very different from those during original construction. Different problems assume different priorities and the primary objective must be to provide structures which can be erected with the minimum overall inconvenience to the public, whether as road user, local resident or consumer of public utility services, and in the shortest possible time.

ACKNOWLEDGEMENTS

49.   The author gratefully acknowledges the Director of the South East Construction Programme Division of the Department of Transport for permission to publish this paper, and wishes to thank his colleagues, in particular Mr. A.G. Simpson and Mr. W. Sim, who were greatly involved throughout this project, for their invaluable assistance.

PART 1: CONSTRUCTION

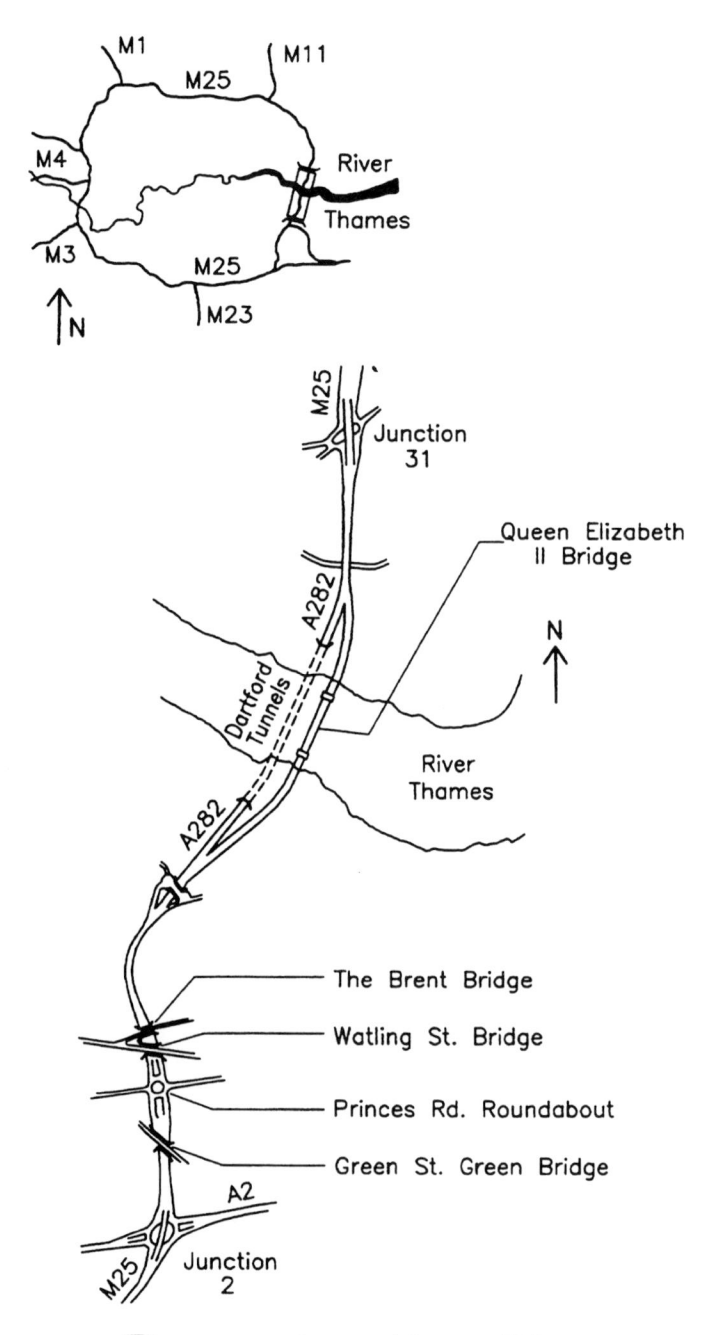

Figure 1. Key Plans

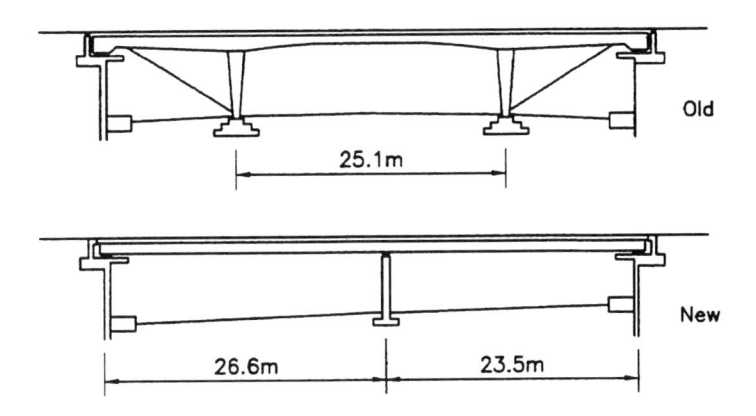

Figure 2. The Brent Bridge

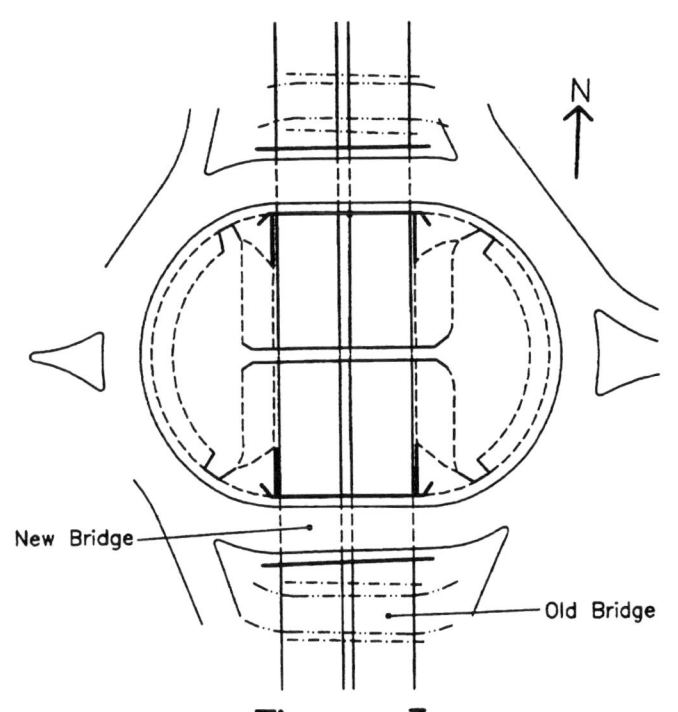

Figure 3.
Princes Road Roundabout

PART 1: CONSTRUCTION

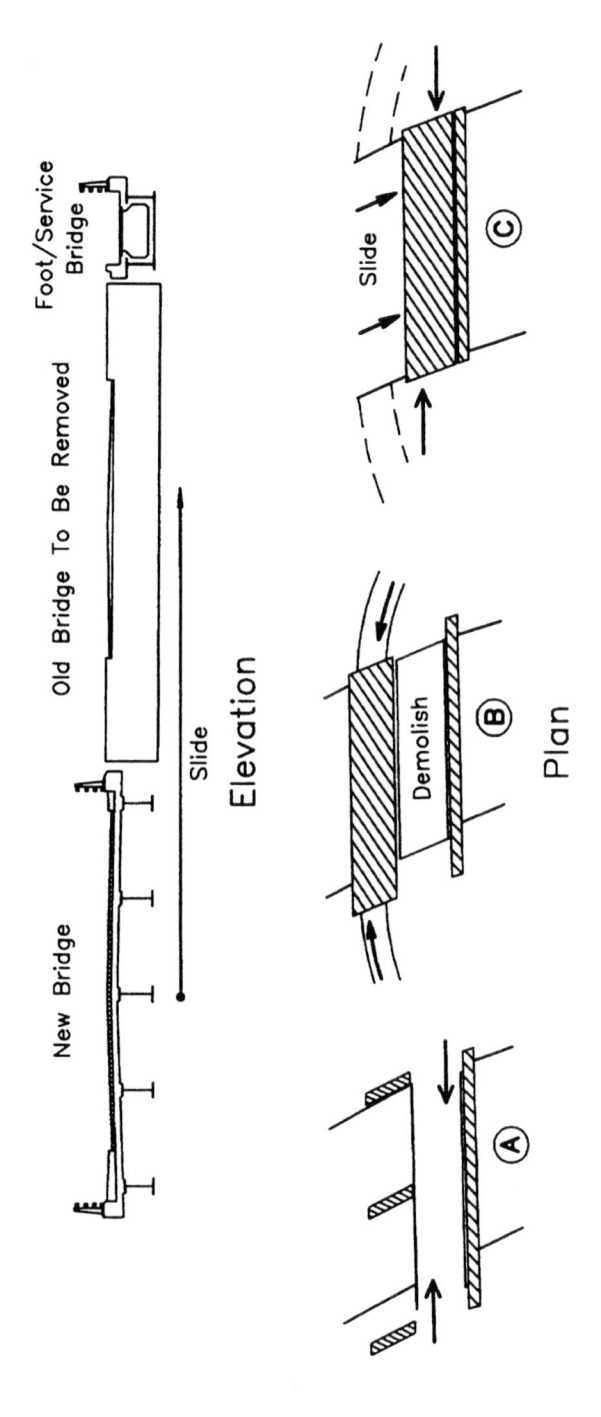

Figure 4. Watling Street

102

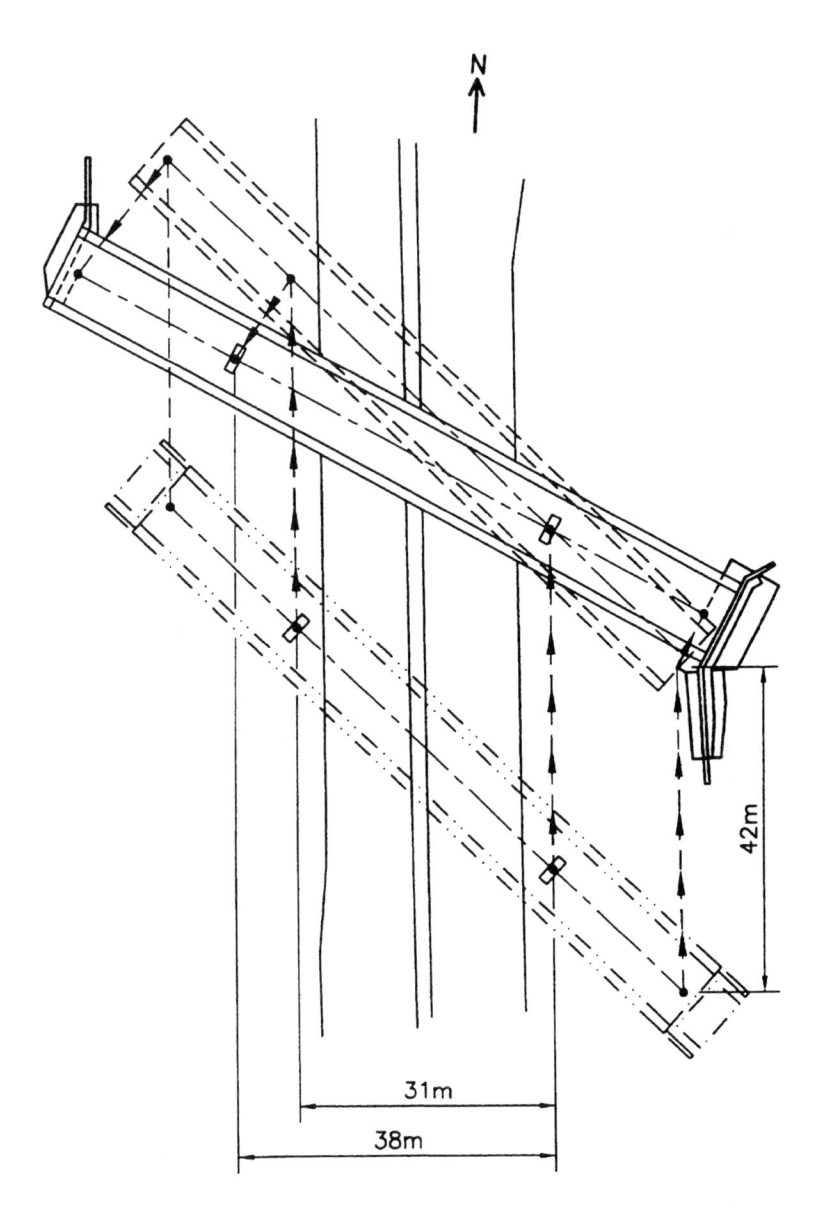

Figure 5.
Green Street Green Road

Widening of small bridges

A. E. NORFOLK, Bridge Manager, and G. D. PETTITT, Kent County Council

SYNOPSIS. With ever changing standards of design and safety, increasing accountability of both bodies corporate and the individual and it seems, the insatiable demands of road traffic, there is a continuous need to improve the existing highway network. In doing this it is desirable to seek to make as much use as possible of the existing infrastructure.

It is in this situation that the subject of bridge widening is considered. The likely scenarios which would involve widening are identified, the methods of achieving that end considered, and the related problems and benefits discussed. The issues are then illustrated by some specific examples.

INTRODUCTION

1. Bridge widening, although not new, is probably more a feature of the 20th century than of any other because of the extraordinary growth in the volume and size of vehicular traffic over its decades.

2. In considering the need for bridge widening it is probably better to broaden the scope to cover highway widening as this allows the full spectrum of situations which are likely to necessitate the extending of bridges both longitudinally and transversely to be identified and addressed.

3. Often these seminars deal with the grandiose and the magnificent, the strategic, specific and the significant. There is obviously a need for these but sometimes the lesser structures, their problems and the broader issues get overlooked.

4. This paper will hopefully address these issues by giving a general résumé of the possibilities for and limitations of bridge widening or lengthening and by illustrating some of them through specific examples.

5. For the purpose of this paper a small bridge is taken to be any of those in the span range from a typical motorway overbridge through the plethora of county highway bridge types to the minimum size for a designated highway structure.

THE REASON FOR WIDENING

6. The reason for bridge widening may be as a consequence of an overall highway scheme or of a particular problem at the bridge itself. The specific circumstances can be summarised as in the following paragraphs.

7. A highway improvement scheme to:-

   a) facilitate an upgrade in strategic importance of the route;
   b) accommodate a local development;
   c) resolve a shortfall in the traffic capacity of the road for its every day (and projected) use;
   d) construct a slip road;
   e) resolve a local visibility problem by improving sight lines;
   f) provide a footway where none exists, due to increasing demand;
   g) provide a hard shoulder alongside a carriageway;
   h) bring a route up to current standards of design and/or safety.

8. An improvement local to the bridge to:-

   a) resolve a visibility problem arising from intrusive parapets;
   b) strengthen sub standard parapets;
   c) remove a restriction to the carriageway due to it narrowing over the bridge or due to wall shyness arising from a lack of verge in front of the parapet;
   d) remove a discontinuity or narrowing of a footway or hardshoulder at a bridge;
   e) accommodate enhanced geometrical standards for improving public safety such as affording necessary clearances and set backs to safety fencing or other obstructions;
   f) allow a more economic form of traffic management to be used on a job than the alternative of temporary bridging, closure and diversion where that job would not, of itself, require the bridge to be widened.

9. Other examples of highway schemes where bridge widening may be involved are:-

   a) reconditioning where a structural haunch effectively widening the carriageway, may result in a bridge parapet becoming more exposed to impact or creating a greater intrusion to vision, with the potential for increasing the risk of accidents;
   b) widening to enable future maintenance on the route or at a bridge to be carried out without having to close the road. The restrictions now on minimum lane widths and safety zones for even the most simple of operations are causing an increase in the number of requests for road closures.

PART 1: WIDENING SMALL AND LARGE

THE WAYS OF WIDENING BRIDGES

10. There are many possible approaches to widening bridges all of which have some merit in particular circumstances. However, widening an existing bridge as opposed to a complete reconstruction is not always the wisest move. Whilst there may be good economical or environmental reasons for doing it, it can equally be a technique where resolution of the problems of achieving a monolithic construction, if required, can be costly, and unsightly. Also structural interaction can cause hidden long term problems if compatibilities of different forms of design and construction and age of materials are not fully addressed or understood.

11. The techniques of widening an underbridge comprise:-

   a) widening the bridge in like construction to that existing;
   b) widening the bridge in a different form to that existing;
   c) overdecking or redecking, with side cantilevers, on existing substructure;
   d) redecking on extended capping beam to abutments (and piers);
   e) construct new bridge alongside;
   f) provide separate footbridge;
   g) revamp lane widths on existing bridge to provide extra lane(s).

12. In the case of widening a highway under a bridge then this can be achieved by effectively lengthening the existing overbridge by:-

   a) providing an extra span;
   b) extending deck to a new abutment;
   c) as for (b) but using techniques such as cable staying or through girders to achieve the extension;
   d) introducing a new lane or hardshoulder under a sidespan.

or by increasing the number lanes within the existing carriageway.

13. The fall back situation in all cases is of course full reconstruction.

THE PROBLEMS AND PITFALLS

14. The potential problems hinted at in paragraph 10 are many and varied. Some apply to all techniques some are specific to one or two. A designer needs to be aware of all of these at the outset to minimise abortive effort.

Widening in Like Construction

15. This generally means construction of foundation, substructure and superstructure over the widened section and it can be either symmetrical or asymmetrical about the existing bridge.

16. In the case of arch bridges this technique does afford the possibility of removing existing facing materials for reuse on the new elevation. Whilst this can be a selling point when seeking to do work on a listed building or in a conservation area, such is the subjective nature of these considerations that the possibility of losing the original structure can also kill the proposal. Recent experiences have found conservation planners objecting to modern materials being used in such hidden situations as to saddle an arch because they believe them to be "an intrusion and a fake" and also employing conservation engineers to back up their policies by proposing more "acceptable" solutions. Conflicts between highway planning, accountability for public safety and conservation are resolvable in an atmosphere of mutual recognition and compromise. In the face of intransigence a stalemate can be reached which does not benefit the highway engineer, the planner, nor ultimately the public they serve.

17. In the case of beam and slab construction the likely variation in characteristics between new and old can be a problem if it is intended to connect the decks structurally. This is especially true the greater the time between construction dates. Clearly consideration must be given to the secondary stresses which are now likely to be experienced by the old deck and what must be done to cope with them. These issues are discussed later. In the event that there is to be no structural connection between the old and the new, then the butt joint must be under an untrafficked part of the highway to minimise the effects of any differential movement.

Widening in a Different Structural Form

18. Again this will mean construction of foundation, substructure and superstructure over the widened section and it can be either symmetrical or asymmetrical. However, in this situation the different characteristics of the sections of deck will be of greater significance making structural interaction between the two more difficult to accommodate. In this event the reference to the location of butt joints in paragraph 17 is especially applicable.

19. It is unlikely that this approach would be acceptable in an area of environmental importance other than with façades incorporating original materials.

20. Some general points which can be made and which apply to widening in both similar and different forms of construction are:-

    a) Generally it will be more economic to widen on one side only. Also this makes for a larger, more stable construction especially if there is to be no structural interaction between the substructures.

    b) Differential settlement between new and old construction is a real possibility and consequently most widened sections tend to be supported on piles. However, this is not a panacea and this issue is discussed further in paragraphs 51 to 55.

PART 1: WIDENING SMALL AND LARGE

    c)   A decision needs to be made as to the design standards to be employed for the widened section of the bridge. It must be an integral part of a scheme such as this that the structural capacity of the original bridge is assessed. The decision will depend on the outcome of this assessment. See paragraph s 43 to 49.

Overdecking or Redecking with Side Cantilevers

21. The essential ingredient of this solution is that the existing substructure is not extended. The original foundations must be capable of accommodating the extra load without detriment to the integrity of the abutments, intermediate piers and original deck if left in place. Obviously, in this respect, an assessment of the substructure is required and, if there are any doubts, remedial action (underpinning and strengthening) must be taken to avoid the likelihood of abortive costs being incurred on the superstructure.

22. Often, with the benefit of modern materials, it can be arranged that the improved deck weighs no more than the original. This will help the engineer greatly in exercising his judgement as to whether or not to improve or replace the existing substructure.

23. One common situation where this type of work is undertaken is on the older arch bridges where the mass of the structure itself generally means that any change in overall weight is not significant thus permitting minimum change to the substructure. This approach also enables the original structure and most of its elevation to remain. This is an attraction to conservationists provided that the overhang of the cantilever is not too dominant.

24. Generally such widening would be symmetrical but care must be exercised in assessing the effects of asymmetric live loads when only one lane is trafficked.

Redecking on an Extended Capping Beam

25. Where it is necessary to remove the existing deck then this offers the opportunity to redeck in conventional form spanning between abutments and/or piers across the full width of the deck with all or some of the widening being accommodated on an extended capping beam cantilevering beyond the edge of the abutment or pier. Further widening can be achieved if required by cantilevering the deck.

26. Otherwise the comments in paragraphs 21 and 22 apply equally here.

Construct a New Parallel Road Bridge or Footbridge

27. In both cases current design standards must be used. It must be borne in mind that along with this it may be necessary to do some strengthening work to the existing bridge if it has failed its assessment.

28. One crucial decision will be whether to butt the bridges or leave a gap between them. If the latter then the gap must be wide enough for access for inspection and maintenance (minimum 1.0m, preferably 1.5m). If butting then the joint must be sealed. This is difficult to do with the inevitable differential vertical movement of the decks. Although separation means changes in alignment of the carriageway and additional runs of parapet, it is generally preferred as all faces are then fully accessible for monitoring and repair. Butted decks are more acceptable if the adjacent vertical faces are shallow as on the end of sloping cantilever parapet beams, so that the edge face of the main body of the deck is accessible.

Provide Extra Capacity by Providing More but Narrower Lanes on Existing Bridge Decks

29. This solution achieves an effective widening rather than an actual widening. It encourages more traffic onto the bridge deck at any one time and the consequent increase in live loading is unlikely to be acceptable when the bridge is assessed. Strengthening may well be needed.

30. This approach also creates great difficulties for inspection and maintenance. There is no room for access without traffic management. Delays when lane closures are necessary are compounded because with safety zones, a minimum 2 lanes are likely to be needed to be closed for the simplest of operations and the larger commercial vehicles will need to be accommodated in the narrower outer lanes.

Overbridge Lengthening

31. Generally this is a much more complex affair than widening affecting, as it does, major elements of substructure and, in some cases, the overall articulation of the bridge. Depending upon the age of the existing bridge, lengthening may not be the best solution for long term benefit.

32. Providing an extra span is probably the least complex of the structural options as it need not affect the overall articulation of the original structure. It can also be designed to current standards as it is effectively an independent structure. Obviously the existing abutment must be adapted to suit. This could be a useful solution for providing a slip road or a continuous hardshoulder but it would involve lane separation if used as part of the main running carriageway and so is generally to be avoided in this last case.

33. Extending the deck by stitching on an extra length to stretch a span has been done. However, it is complex and does not offer long term benefit as the original design standards will govern the capacity of the overall bridge unless it is feasible to strengthen the remainder of the bridge at the same time.

34. The options involving cable staying or through-girders to support and extend the bridge overcome the design standard problem as they offer enhanced structural capacity. The latter also presents the extra benefit of removing intermediate piers

if so desired. However, there is a clear penalty to be paid in aesthetics with the new structures being much more visually intrusive. Also the economics of these types of solution should be compared with that of the more satisfactory new bridge. Avoiding disruption to Utilities is an obvious advantage although it is argued later that these should not be in the economic equation.

35. In all these lengthening options achieving headroom can be a serious problem as bridges are often dipping or sloping down to an abutment and extending this alignment will aggravate the situation. Also the options in paragraph 34 will intrude into the existing clearance beneath the deck. It will therefore be necessary to consider the possible effects of carriageway lowering and/or deck raising in the cost comparisons.

36. Providing a new lane under an existing sidespan avoids affecting the bridge in any way but may require a retaining wall to support the ground in front of the abutment. In this event the requisite safety measures for the inspector when looking at the bearing shelf and revetment area must be installed. Also the width of carriageway provided must be adequate to allow its maintenance and that of the wall and the abutment to proceed without closure.

37. As mentioned in para 32 this also could be a useful solution for a slip road or for providing a hardshoulder but would again involve lane separation if used as part of the main running lanes and so is generally to be avoided in this case.

38. This technique has also been used to provide a high load clearance under a side span, the new lane being lowered relative to the existing carriageway.

Increasing Capacity of Existing Carriageway under a Bridge

39. Increasing the number of lanes within the existing carriageway generally means loss of any existing hardshoulder or strip. This removes a major benefit in respect of the options available for traffic management and creates hazards during such works as some of the lanes are generally narrower than normal standard. Furthermore, if new discontinuous hardshoulders are provided then the intruding piers or abutments present an additional hazard to motorists.

Other Potential Problem Areas

40. Public Utilities are always a major problem area in highway works and widening schemes are no exception. It would be wrong to presume that there is no problem if the existing deck is unaffected. If the utilities were in the verge originally and the bridge is widened they will now be in the carriageway with all the attendant problems that presents for maintenance. Similarly, if they are attached to the outside of the existing bridge they will need to be accommodated in the widened structure. There is a benefit here as the original eyesore will be removed. Perhaps consideration can be given to providing a separate crossing for the utilities and then declaring the main structure to be protected. Either way a diversion is necessary, adding to the duration and cost of the job. Further comments are made in paras 63 and 64.

41. Lack of details of the existing structure can present all manner of problems. In particular, the less that is known about a bridge the more costly the investigation will be to uncover the facts or the greater the risk will be of finding something wrong with the old bridge during the contract. This issue is discussed in more detail in paras 65 to 68.

42. Other influences on the choice of solution are:-

   a) Type and duration of traffic management needed to facilitate the work.
   b) Nature of what is being spanned and related accessibility.
   c) Condition of existing bridge.

DISCUSSION ON SOME OF THE PROBLEM AREAS

Changing Design Standards

43. It is probably inevitable that, in widening an existing bridge, the engineer will seek to carry out the design to current standards. The approach to be used in dealing with the existing bridge will depend upon the type and scale of the widening.

44. If the widening is significant and all on one side then it is possible to accept a new and old concept, always presuming that the original structure has passed its assessment and has adequate provision for abnormal loads. In the event that the original deck has failed to meet these requirements then either the deck can be removed and a whole new deck be provided or the existing deck can be strengthened to either the assessment load or current design standard, preferably the latter.

45. If the widening is symmetric then the logic for differing standards of deck design diminishes and strengthening or redecking become much more likely.

46. One alternative is to design the extension to the same criteria as the original. This would be difficult to sustain in the light of a Highway Authority's accountability for public safety. Although there may be a temptation to do this in quiet rural areas such a location seems at the same time incompatible with the need for widening. Nor can there be certainty over what traffic may use a route in the future especially if there is a need at any time to close an important neighbouring route.

47. Changing design standards generally reflect a growing awareness of how structures and materials react to everyday and long term use, shortcomings in previous standards and a growing vehicle stock in both size and number. They also challenge the philosophies and sensitivities of engineers who are now charged with 'condemning' bridges they have designed and built. In the climate of accountability and risk which now exists, engineering judgement may be considered a liability, and certainly there is no tolerance allowed today in exercising it. However it still must play an important part in determining these situations.

48. In such cases the designer must sometimes consider that, whilst the old is to all appearances sound and fit to continue in use, there is not some fundamental illogicality in adding to it bridge width which has been designed with the benefit of all the latest in technology. He will be well aware that the old section was probably designed to no particular standard, or at best to a simply-applied strip analysis. Perhaps such design as was applied was left to the mason who put it up on the basis that if the arch stayed up when the backfill was placed then it would be quite satisfactory for the occasional horse and cart. There are even more troublesome choices. These are between the merits of a 1960's design, when the belief of the engineer in his ability was absolute and relatively few academic authorities' were expounding on design matters, and those of the more sophisticated and more lengthy design rules which have followed in the eighties and nineties.

49. Where the widening is of an arch by a further arch the engineer can perhaps convince himself that, since design rules for masonry arches still do not carry a great deal of academic weight, it remains acceptable for him to use his experience and judgement in detailing his widened bridge. Where, however, his materials are reinforced or prestressed concrete or steel composite, he will feel less confident that what already exists is fit to be allowed to abut the latest in bridge technology. Indeed, it is often the case that an apparently perfectly functional bridge is effectively discounted from continuing in use by the safety-first philosophy that requires any bridge deck which fails its assessment to be strengthened or replaced to the most recent design standards.

Varying Characteristics

50. Varying characteristics, both structural and material, coupled with changing design standards pose particular problems when seeking to secure structural interaction between new and old construction. Careful consideration must be given to the transverse distribution and torsional effects at the joint and over the old deck. Two specific scenarios exist:

    a)   Arch Bridges

    i)   Brickwork and masonry arches have been widened using reinforced concrete fill and overslabbing. The mode of action of an arch with blocks and joints is quite different to that of a reinforced concrete deck slab. One can imagine, and one is encouraged to believe, that the multi-element arch is flexible and can adjust itself to accept live loads and temperature movements throughout its whole: barrel, spandrel, parapets and foundation, whilst the slab can only deflect due to load in a vertical direction and expand and contract in the plane of the slab. To sit the slab on a cut-down arch or fill the arch with rigid unyielding reinforced concrete can be visualised as a sure and certain way to produce problems associated with incompatibility of stiffness and indeed such incompatibility is as much a function of shape and mode of action as of difference in materials.

ii) However, steps can be taken to separate materials with thin membranes so that they can continue to act independently, and therefore flexibly. Alternatively, the two materials can be so linked that they act homogeneously. There are undoubtedly theoretical problems which are created by linking one material with another but it has been done quite successfully for many years and saddling of arches in reinforced concrete is a good example of this process. Fortunately coefficients of expansion are not wildly different and where interfaces of different materials do occur they are unlikely to be subject to vast changes of temperature.

b) Beam and/or Slab Bridges

i) The problem which arises when it is proposed to widen a deck composed of beams and slab between, be they steel or concrete beams, is the likely incompatibility of stiffness between the old and the new. It is suggested that for a long span both the old and the new decks can be idealised and subjected to live loading in some form of modern computer analysis. In this way whatever differences of stiffness exist can be simulated and can be accommodated in the design and detail of the new deck, of the connection between the old and the new and, if necessary, in the strengthening of the old.

ii) For a short span, if the new and the old deck are substantially of the same form of construction, the differences in deflection will be small but there will nevertheless be a tendency for the stiffer side of the connection to be left supporting the less stiff side. If this less stiff side is the old existing side it is likely that it will see less design load than it should and than it is capable of taking. But the new side can be so designed that it can accept the extra transmitted load thus making the joint satisfactory. The reverse is, however, not true and the transfer of extra load from the new into the old would not be acceptable unless it is economical to strengthen the old deck to accommodate the increased stres

iii) Alternatively, the joint between the new and the old can be a butt joint without structural connection. As noted previously, this joint would probably be acceptable in a verge or a central reserve if the edge face is shallow and suitably sealed, but it is not acceptable in the trafficked lane of the carriageway.

Settlement and/or Differential Settlement

51. Sometimes it is possible to widen a deck physically without widening the substructure. Sometimes the substructure too must be widened. In either case it is desirable to know to what size and shape and of what material the existing foundation is constructed. Cores can be taken through abutments to establish thicknesses and material properties. Trial pits can be excavated in front of, but

rarely behind, abutments to establish depths and widths. Almost inevitably one will find, on carrying out an analysis, that the abutment will not stand under modern loading conditions yet it has been standing there for scores of years! What cannot often be determined is whether or not there are timber piles, rafts or mattresses under the foundation and even if these can be identified it is difficult to evaluate their condition and effectiveness.

52. It has been the case, given that an existing foundation is manifestly working well, that the design engineer can convince himself, and indeed his client, that as long as his new deck or his widened deck plus what remains of the existing deck does not exceed the weight of the original, then the foundation will be satisfactory for the foreseeable future. As mentioned previously significant savings in weight can often be made by for example replacing stone or brick parapets by steel although for environmental and appearance reasons this may be unacceptable.

53. The soil beneath an existing foundation will have been compacted and consolidated. The likelihood is that any settlement that was due to take place on that soil under that load will already have happened and that there will be little more to come. The soil under the new foundation will not have been subjected to quite the same forces. When it does become part of the new structure and receives its share of the loading, it will probably settle relative to the existing.. The effects of this exhibit themselves simply as a difference in the carriageway levels identified by cracking in the surfacing and ultimately a longitudinal step in the carriageway. But, as already discussed for the deck, it could also be characterised by the transfer of loads from the new to the old part of the structure with the consequence that neither the new nor the old foundation is acting as idealised.

54. One way of overcoming this problem is to pile the new foundation and thus eliminate, as far as possible, most of the settlement under the new. The drawback with this is that the final combined structure is left for evermore with an inconsistent style of foundation which might, over the years, react differently to ground movements, changes in ground water level or any other possible cause that would be better served by a foundation of the same design.

55. In order to obtain consistency, piling can be carried out through the existing foundations. This is, of course, expensive, and the designer will begin to wonder whether there is any virtue at all in retaining any of the existing structure. Sometimes when great doubt exists as to the adequacy of an abutment it is useful to build a new substructure behind the existing so as to avoid the need for significant temporary works. This saving would need to be compared with the cost of providing the extra length of span required.

Planning Constraints

56. These are an increasing problem. Power without accountability is a dangerous thing and this seems to be a situation which is developing in some areas in conservation planning.

57. It must be stressed that many such planners are able to discuss all potential options in a constructive way and also accept that others have responsibilities to the public, most importantly for their safety on the highway. Some however see only their own policies to the exclusion of all other interests, a trap which bridge engineers must avoid. The comments in para 16 are particularly relevant here.

58. What is evident is that an early approach must be made to the local Planning Authority to discuss the job and its particular problems to secure co-operation from the outset. It is useful to consult a planning specialist and obtain his advice in the first instance. There are no simple rules to follow here as much of the judgement is subjective. However if a willingness to be seeking to achieve as much as possible of the planning requirements can be demonstrated then, in most cases, an acceptable compromise can be reached.

59. It is certain that the external features and the parapets will be the area of greatest conflict but two worrying trends are those which seek to refuse consent to anything which alters the structure in its fullest sense, even if it will not be seen, and the employment of environmental engineers to back the conservation case. Their advice is given without the constraints of any accountability at all for standards of public safety.

Inspection and Maintenance

60. In carrying out a widening scheme due thought must be given to access for inspection and maintenance.

61. In this respect the issue raised earlier of seeking to increase the number of lanes within the existing carriageway, generally by losing a continuous hardshoulder, is a disaster. Where access was difficult before this idea simply creates more havoc when operational. Where previously it might have been necessary to have only one lane out of two or three closed to work on a bridge, in the changed circumstances it is likely that two out of four, will be required with the heavier vehicles being pushed into the narrower outer lanes. This will result in worse congestion than ever. Such an approach is very short sighted and ignores all the lessons which should be learnt from past experiences.

62. Also in butting two decks together thought must be given to access for maintenance of the adjacent decks and sealing of the joint. In reality it is probably better not to let the decks touch but to have a small clear gap under the seal to avoid entrapment of detritus and potential permanent dampness. Keeping the depth of the face as shallow as possible as mentioned in para 28 is important.

Utilities

63. The cost of Utilities works can have a huge influence on the cost of a job and in one case recently materially affected the outcome in a repair or replace assessment. It is possible that such costs should not be allowed to influence a decision to the long term disbenefit of the travelling public. The nation has accepted that Utilities can be accommodated in the highway and it should be

prepared to accept the costs of such a policy. Utility costs should not be part of the economic equation although this could only be a practical proposal if separate funding was available for them. It is interesting to speculate what will be the situation in the case of future privately built and maintained highways.

64. As mentioned in para 40 widening works, especially where the existing deck remains untouched, are not necessarily free of Utility costs. Where they were originally in the verge they might now be in the carriageway. It would be a disadvantage to leave them there as any maintenance of them will affect the carriageway rather than the verge. However the decision to move them will undoubtedly weigh upon their actual location in relation to available working space and safety zones.

Hidden Difficulties with Old Bridges

65. Old bridge drawings are very often not totally accurate. Our predecessors, even quite recently, did not have the benefits of computer-aided draughting or the rigorous document control requirements of a Quality Assurance System to assist them in updating the status of the record drawings. This, together perhaps with many years of accident repairs, maintenance revisions, strengthenings and the attentions of the utility companies may mean that the designer of a widening scheme is not fully aware of what is the precise situation with the part of the structure that is to be retained.

66. Experience has shown that it is often very difficult to retain parts of an old structure once construction works are commenced. Narrow arch bridges which are due to be widened on one or both sides have been known to give up the struggle when their spandrels are removed to facilitate the works. It is not only the fact that the structures were not formally designed and their stability depends on the time-proven whole structure but that the use of modern machinery sometimes lacks the necessary touch and delicacy such work requires. Hand work can be carried out but economics may rule against this.

67. Recent nationwide concern has concentrated on the condition of postensioned tendons and ducts. There are guidelines for the investigation of these elements but it is clear that this can only give an indication of the condition of the complete structure and it may only be when the greater accessibility offered by the main works is available that the true extent of any problems can be better seen and evaluated. With the contractor anxious to proceed, decisions to retain existing elements may need to be sacrificed to expediency and economics.

68. This sort of problem can lead on to the question of how much site investigation should be ordered before the works are designed. The view sometimes taken is that it is undesirable to add to the inconvenience of the road user by carrying out site investigation on the existing structure when the road is going to be disrupted later by the widening works. This is because it is during the works themselves that the better opportunity arises to view the existing condition of the bridge and to verify its capacity to continue in service. The problem is that

such a late appraisal can have serious consequences for the contract and the period of disruption to traffic if major problems are encountered. Whilst a preliminary investigation cannot guarantee that problems will be avoided entirely, on balance, and with careful planning by experienced engineers, it must pay dividends in most cases and is therefore to be recommended

EXAMPLES OF BRIDGE WIDENING

69. So far, various problems and their solutions have been discussed. To some extent it may seem that there are more problems and pitfalls than there are solutions. The following examples will hopefully demonstrate that the difficulties can be resolved.

Stile Bridge (See Appendix A)

70. This bridge originally comprised three simply supported spans, of 4.5, 8.5 and 4.5m, made up of longitudinal cast-iron beams, with brick jack arches spanning between them, on top of which was a road construction of compacted fill and tarmac. Assessments had revealed that the decks would require the application of a weight restriction which would be unacceptable on the class A road which was carried. At the same time an improvement in width to provide a 7.3m carriageway, two 1.5m footways and clearance to the parapets was sought.

71. The decision was taken to redeck the bridge with a continuous reinforced concrete slab supported on the existing substructure. No site investigation of the substructure was undertaken except for a visual inspection which revealed no evidence of settlement. The basis of the decision in favour of retaining the existing substructure was that the weight of the new deck was to be less than the old. In simple terms this consideration was satisfactory. The problem which was introduced by this decision was that the articulation of the new deck applied a horizontal force due to braking which, in theory, the existing brick abutments would not be able to accept. The solution to this problem was to take this load via an approach slab into the fill and road construction behind the abutments. This extension slab also overcame the problems of achieving the widened bridge over the wing wall area.

Chevenev Bridge (Appendix B)

72. This 8.0m single span bridge comprised a narrow 3.6m wide deck of haunched cast-iron beams and transverse cast-iron plates which was subject to a 3 tonne weight limit. In normal times the weight limit and the width were adequate for its usage but when maintenance had to be carried out on a neighbouring bridge and light traffic was diverted over Cheveney bridge, the limitations posed by the width and weight restrictions became most evident. The greater volume of traffic created passing problems and the weight limit was also abused since the necessary diversion for heavier vehicles was considerably greater in distance.

73. The decision was taken to replace the deck thus facilitating the removal of the weight limit at the site and allowing an increase in width to be provided at the

PART 1: WIDENING SMALL AND LARGE

same time. The solution adopted here was to retain the existing foundations for appearance but, because the new prestressed concrete beam deck was to be heavier than the existing, new spread foundations were provided in the line of the carriageway far enough behind the existing brick abutments not to overload them with the new bridge. All work was able to be carried out at road level since the new foundation was sized to reduce the applied bearing pressure to a minimal level with some of the extra width being provided by cantilevers off the main body of the deck. Generally, this was an economic and a satisfactory solution to a by no means unique problem. There was, however, one site incident which detracted from its general success. In excavating for the new foundation the machine pushed over into the river the top of one of the abutments which had been so carefully planned to be retained for appearance sake resulting in extra works to rebuild it. There also remains a fundamental snag with the way in which this example of longitudinal overspanning was achieved in that, as detailed and constructed, it is now practically impossible to inspect the bearings. This is something which more recent office practice would not permit.

Wye Bridge (Appendix C)

74. This bridge has not yet been widened and the drawing presented shows what seems to be the most acceptable solution to the multitude of interested parties concerned with this Grade II* Listed Building. The basic bridge is a five span masonry arch bridge in ragstone. In the 1960's the very narrow bridge was widened using steel beams inserted as cantilevers part-supported on steel brackets and trimmed up longitudinally with a deep longitudinal Universal Beam. This sounds like a mess and it certainly does not appear any better. It was this steel widening, providing a width of carriageway, a footpath and verge on the bridge, which recently failed a load assessment. (See Appendix C1).

75. The proposal (See Appendix C2) indicates that the required width is now provided by a reinforced concrete slab supported on the strengthened arch construction. The 1960's steel additions have gone. The depth of the new widened construction is less than the old and the existing abutments and piers are preserved. The opportunity may be taken to hide, either in shadow or, perhaps, behind some form of acceptable masking provision, the wealth of services which have been attached to the stone elevation on both sides of the bridge over the years. The cost of moving them is prohibitive.

76. Many alternative proposals have been considered: a half-bridge alongside in modern construction; a full-width bridge in modern construction alongside, setting the existing aside for pedestrians only; widening the bridge both symmetrically and asymmetrically by moving the existing masonry elevations outward and constructing a new matching core in modern arch construction; constructing a new access into Wye altogether enabling the old bridge to be restored at its old original narrow width.

77. There is, naturally, a wide public interest in the solution at this site which provides the major access into the small town of Wye and which is significantly constricted by a railway level crossing and a number of other listed buildings.

Motorway Underbridge Widening (Appendix D)

78. This widening solution differs from the other three examples in that it deals with the widening of a Motorway bridge and it was not designed by the County Council's staff.

79. The need has arisen to widen a dual two lane motorway to a dual four-lane motorway. The existing bridge has been retained and new bridges have been added each side of the existing to accommodate the extra width. The new bridges have been piled whereas the existing was not. From the drawing it is clear that there is no positive structural connection between the new and the old decks. The designer has provided piles to minimise the potential for differential settlement between the old and the new and the butt-joint has been made non-structural to avoid changing the stresses in the existing bridge. This sort of joint was thought possible here because they both occur in normally untrafficked areas of the deck under noses at the slip roads. The drainage detail has been added to prevent drips falling onto the public highway below.

CONCLUSION

80. The considerations involved in any scheme which requires a wider bridge than exists at present are many and varied. It would be unwise to attempt to make any general rules to apply other than the one to judge each site on its merits. In broad terms the issues to be reviewed fall under the headings of condition and strength of the existing structure, the extent and location of widening required and incidentals. Perhaps the most relevant question relates to the residual life of the existing bridge and the difficulties its premature replacement would cause on the widened highway. Retaining an existing structure is not automatically the right thing to do, nor is knocking it down. But it is very necessary to consider the implications for the future of a decision to keep it. Hopefully this paper will assist in this process.

81. Assuming that the old can safely be retained, it is probably better not to mix materials or span arrangements with the extension. It will be much more likely that compatibility of deflection, settlement and expansion are achieved if, as far as possible, widening is carried out like for like. It is not unusual however, to see arches that have been widened using cast iron or steel beams in some sort of composite action with ordinary fill, or using reinforced concrete and working quite happily. It does seem much less likely however that such solutions would be acceptable today not least because of the strong planning and conservation lobby which exists to keep the more extravagant widening schemes firmly in check. Good bridge Engineers will naturally be conservationists at heart but not to the exclusion of all other issues.

ACKNOWLEDGEMENTS:

The authors acknowledge the help and encouragement of Allan Mowatt, the Director of Highways and Transportation, Kent County Council, in the preparation of this paper. The views expressed are those of the Authors and are not necessarily those of the Kent County Council.

PART 1: WIDENING SMALL AND LARGE

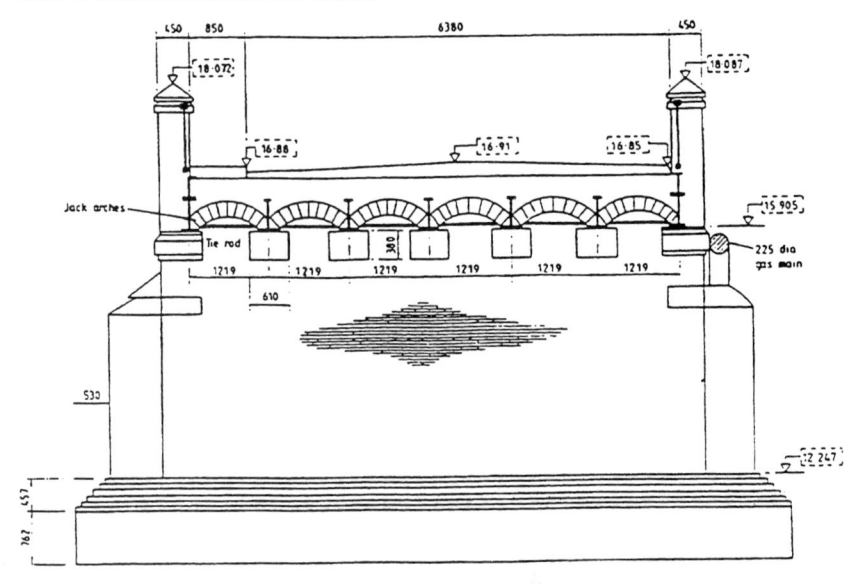

SECTIONAL ELEVATION A-A

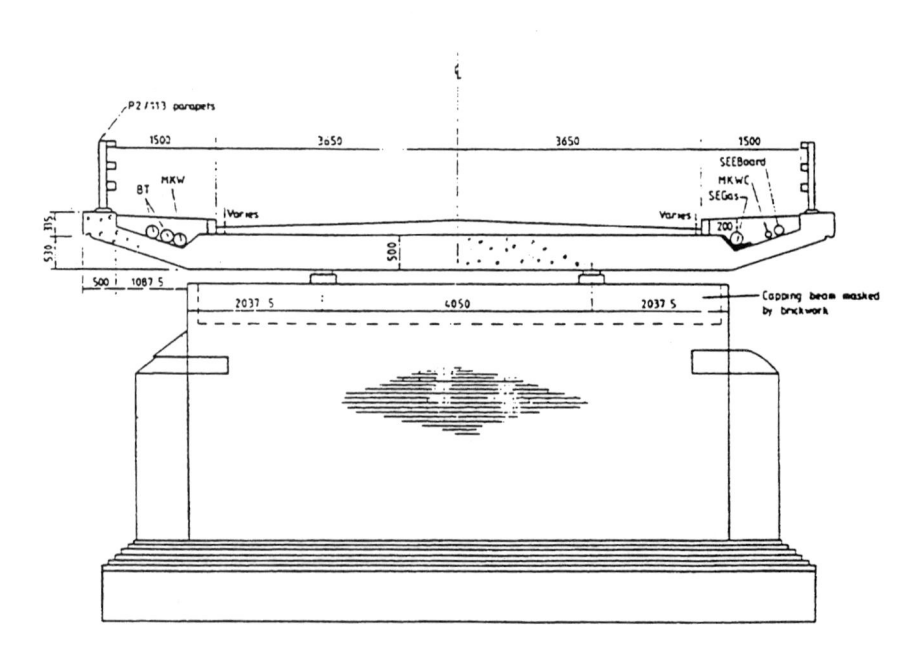

SECTIONAL ELEVATION D-D
Scale 1:50

APPENDIX A - STILE BRIDGE No. 67

Showing:-
* (top) Old Cross Section
* (bottom) New Cross Section at retained Existing Pier

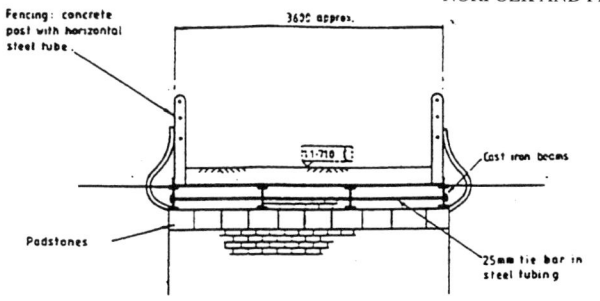

CROSS SECTION THROUGH EXISTING DECK
SCALE 1:50

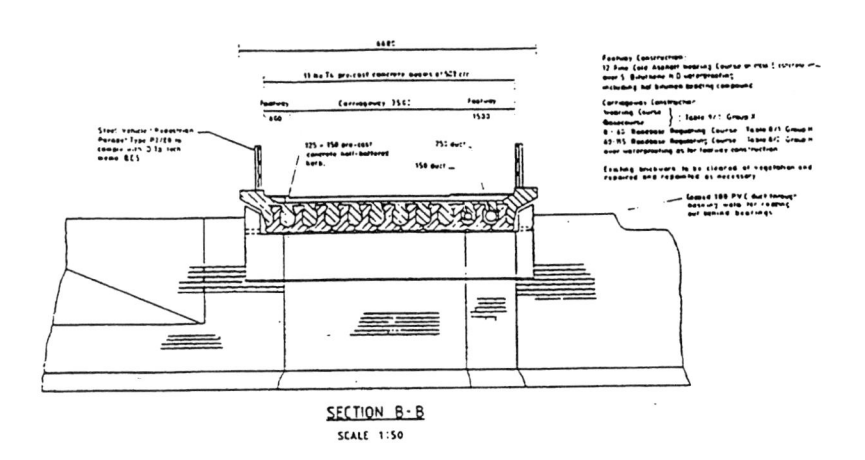

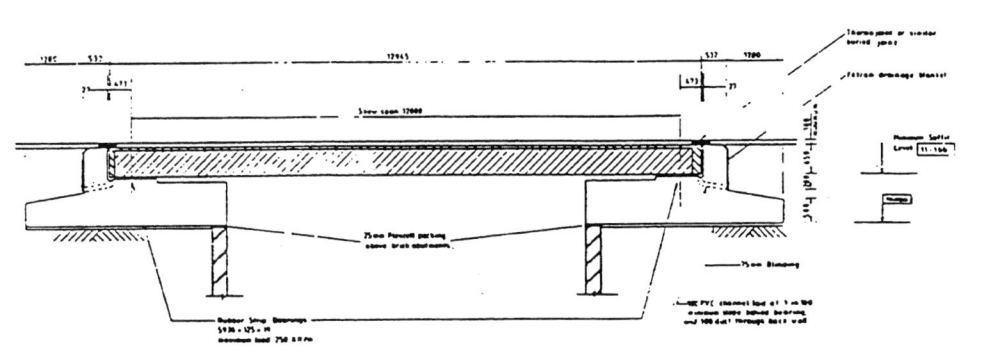

SECTION B-B
SCALE 1:50

APPENDIX B - CHEVENEY BRIDGE No. 476

Showing (top to bottom):-
- Old Cross Section;
- New Cross Section and the retained Existing Abutments and Wingwalls
- New Long Section showing overspanning New Deck and New Abutments constructed behind the Old.

PART 1: WIDENING SMALL AND LARGE

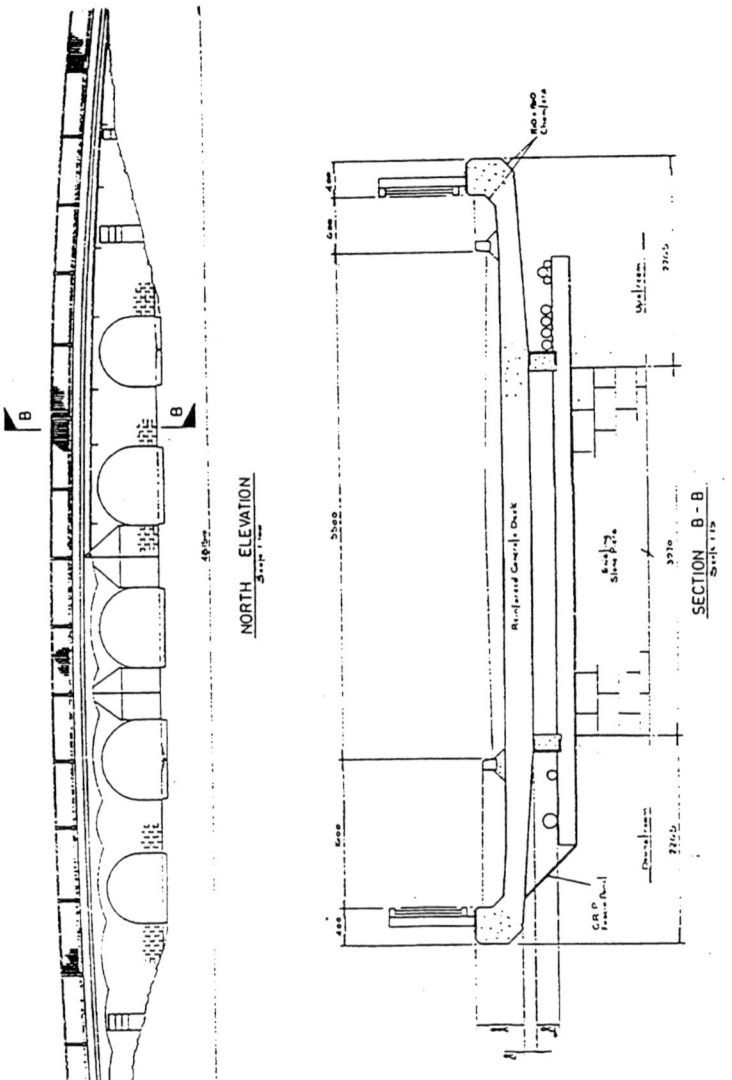

APPENDIX C1 - WYE BRIDGE No. 76

Showing:-
* (top) Proposed Elevation left-hand end with services masked, right hand end with surfaces left exposed
* (bottom) Proposed Cross Section at Pier left-hand side with services masked, right-hand side with services left exposed

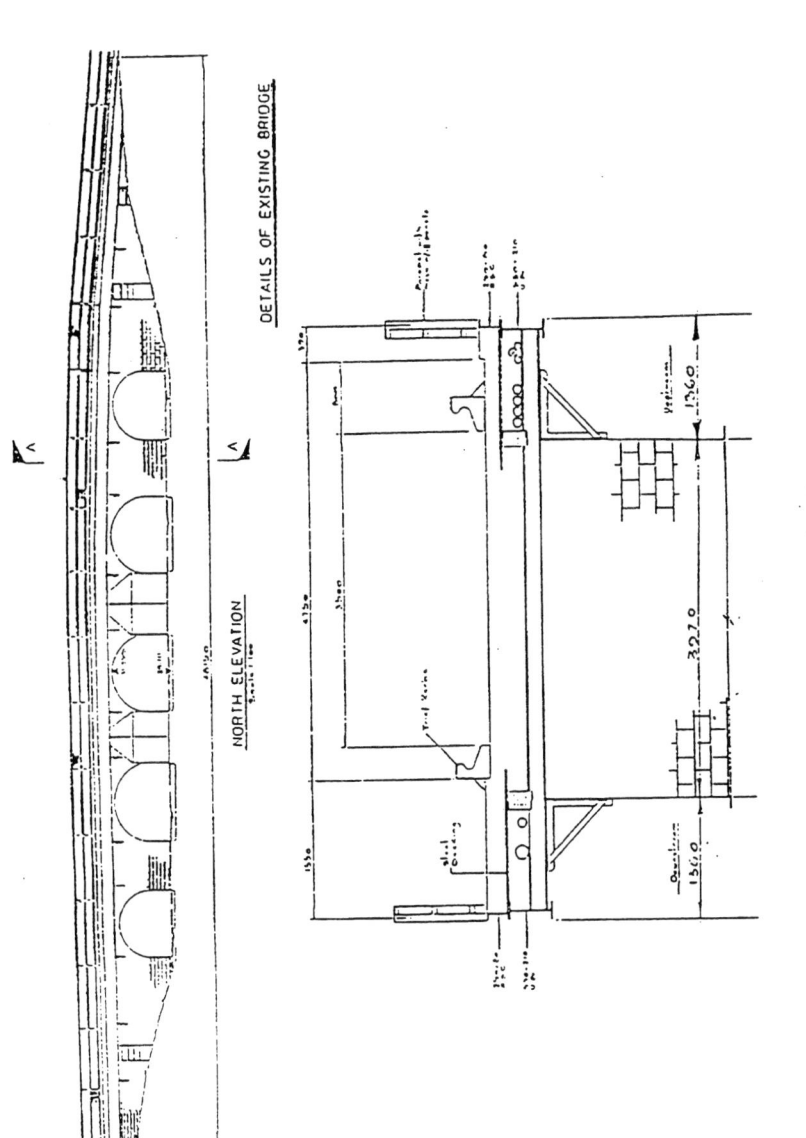

APPENDIX C2 - WYE BRIDGE No. 76

Showing:-
* (top) Existing Elevation
* (bottom) Existing Cross Section at Pier

PART 1: WIDENING SMALL AND LARGE

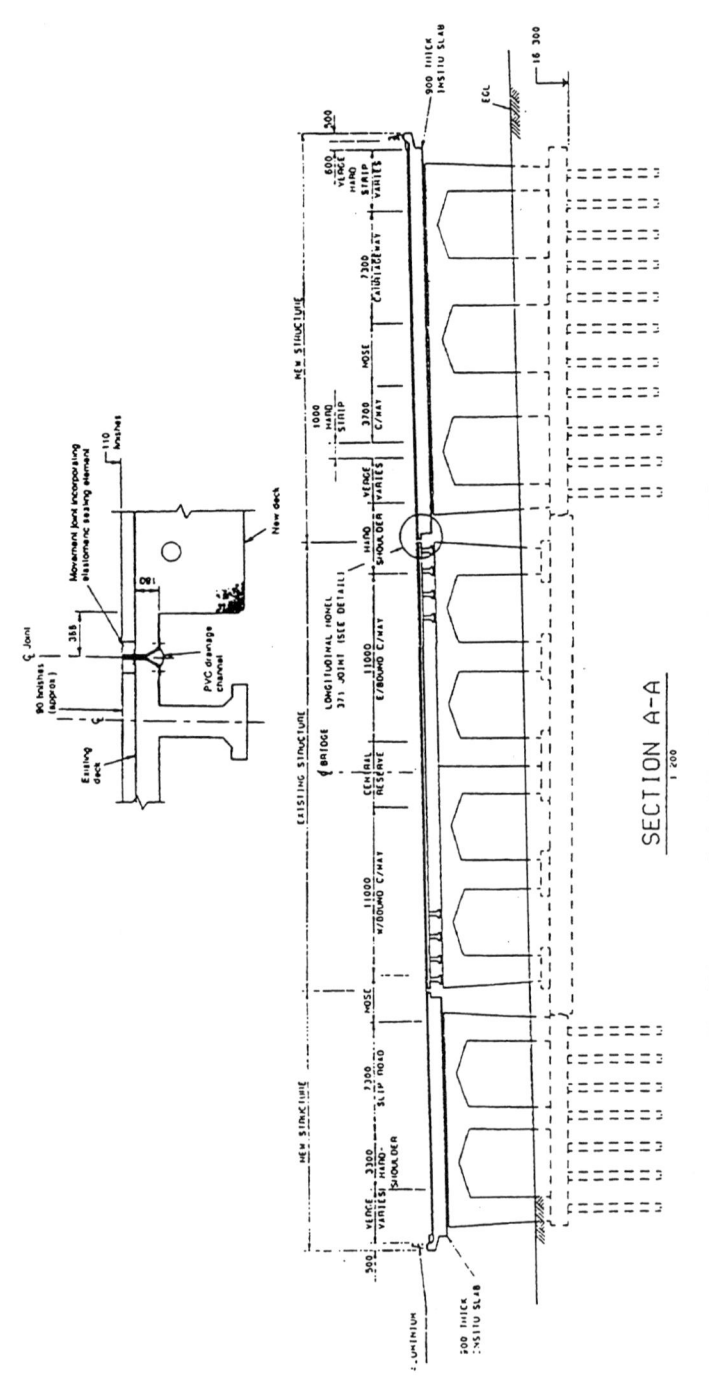

APPENDIX D - MOTORWAY UNDERBRIDGE WIDENING

Showing New widened Cross Section at Pier and enlargement of joint detail

Rodenkirchen suspension bridge reconstruction and widening

R. HORNBY, Senior Bridge Engineer, Rendel Palmer and Tritton

SYNOPSIS. The reconstruction and widening of the suspension bridge over the Rhein at Rodenkirchen in Germany has been under way for three years. This paper reports on the development by Rendel Palmer & Tritton of the client's preferred reconstruction scheme which, when completed will double the width of the existing bridge together with replacement of the existing corroded concrete deck slab. During reconstruction four lanes of traffic will be maintained on the bridge.

INTRODUCTION
1. The suspension bridge carrying the A4 Autobahn across the Rhein (Rhine) at Rodenkirchen, approximately 5km south of Köln (Cologne) was originally constructed in 1938-40. This original structure had a short life as it was damaged by bombs in 1944 and then was finally destroyed by bombing on 28 January 1945.
2. The deck and the cables were completely destroyed, but both towers remained standing, although damaged and slightly distorted during the collapse of the rest of the structure. The cable anchorages were substantially undamaged. The bridge was subsequently rebuilt between 1952 and 1954 to a design by Dr.Ing. H. Homberg which retained the towers and anchorages.
3. In 1990, a contract was awarded to a consortium consisting of Strabag Bau AG, Thyssen Engineering Gmbh and Cleveland Bridge and Engineering Ltd. for further reconstruction and also widening of the bridge. The consortium appointed Rendel Palmer & Tritton as their consulting engineers responsible for the final design of the widened and reconstructed bridge superstructure and also for the analytical work associated with the erection of the bridge superstructure.

PART 1: WIDENING SMALL AND LARGE

DESCRIPTION OF EXISTING BRIDGE

4. The reconstruction of the bridge in 1952-54 utilised the existing substructure and towers. No remedial work was required at the anchorages, but one of the piers had suffered some cracking damage which required repair. After straightening and repair the existing towers were judged to be capable of re-use. The 1952-54 reconstruction therefore essentially consisted of the addition of new cables and a deck structure onto the existing 1938-40 substructures and towers.

5. The existing bridge structure is a conventional three span suspension bridge with principal dimensions as shown in Fig.1 below.

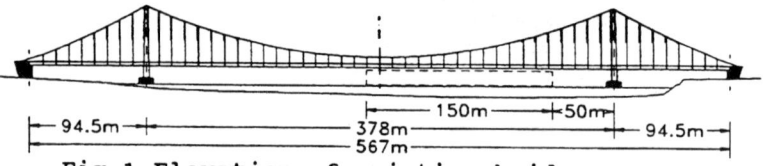

Fig.1 Elevation of existing bridge.

6. The deck cross section is as shown in Fig.2. Plate girder stiffening girders 3.3m deep support cross girders at 2.625m centres. These in turn support longitudinal stringer beams which act compositely with a concrete deck slab. The deck slab is in general 190mm thick but is thickened locally to 290mm near the towers. The slab is post tensioned both longitudinally and transversely.

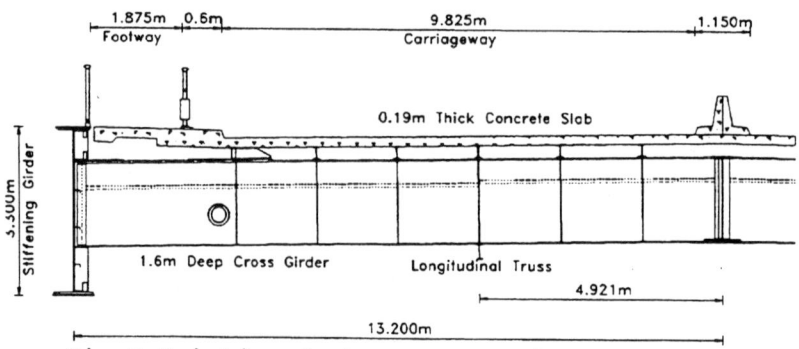

Fig.2 Existing deck cross section.

7. The existing towers (shown in Fig.3) are of rivetted box construction and support main cables each made up of 61 locked coil ropes of 54mm dia. arranged in a hexagonal formation. From these

cables the bridge is suspended by 60mm dia. wire rope hangers at 10.5m centres.

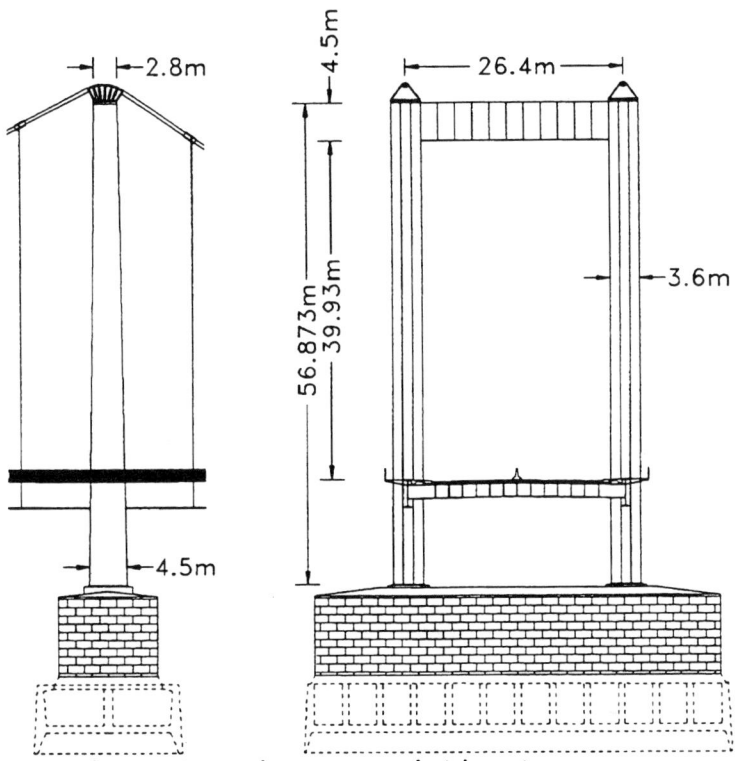

Fig.3 Elevations on existing towers

REASONS FOR BRIDGE RECONSTRUCTION

8. The concrete deck of the 1952-54 reconstruction was designed without any asphalt surfacing. As a result of the usage of salt for road de-icing, extensive deterioration of the road surface had occurred by the early 1970s. In an attempt to counteract this deterioration, an asphalt surfacing layer was added to the bridge in 1975. However, later investigations showed that the corrosion of the deck reinforcement was continuing and that this would eventually require replacement of the concrete deck.

9. It had been decided to widen the A4 autobahn from two to three lanes in each direction. It was therefore necessary to provide additional traffic capacity across the bridge, either by replacement or widening.

DESIGN ALTERNATIVES AND CHOSEN SOLUTION

10. The initial investigation into possible methods of reconstructing the bridge deck and increasing its capacity to match that of the widened A4 autobahn was carried out by the Engineering Department of the Landshaftsverband Rheinland (LVR). A number of options were considered, including:

(a) replacement of the existing concrete deck with a new steel orthotropic deck capable of accommodating 6 traffic lanes of reduced width;

(b) replacement of the existing concrete deck by a steel deck with the addition of an upper deck;

(c) reconstruction of the existing deck combined with the construction of an immersed tube tunnel alongside the bridge;

(d) replacement of the existing bridge by a new cable stayed bridge carrying six traffic lanes; and

(e) construction of an identical suspension bridge alongside, combined with replacement of the deck slab of the existing bridge.

11. All of these options proved to be unsatisfactory for either structural, operational, alignment or land-take requirement reasons. LVR therefore turned its attention to the possibility of widening the existing bridge. It was clear that even with the reduction in dead load provided by replacement of the concrete deck with one of steel, additional cable capacity would be required. The simplest way in which this can be achieved is by the provision of one additional cable, and combined with widening of the deck on one side to double its existing width, the symmetry of the bridge is preserved.

12. For this scheme to work a means of redistributing the additional loading away from the cable at the centre of the widened structure is required. The most convenient means of achieving this is by pulling in the new cable at each of the anchorages to effectively shorten the new cable. The deck transverse stiffness then redistributes the loading between all of the cables.

13. This scheme has major advantages over the others considered, namely:

(a) minimising alignment changes to the autobahn on either bank, and whilst at the same time minimising land-take requirements;

(b) preserving the outline and general appearance of the bridge, which is regarded locally as an important landmark.

This scheme (as shown in Figures 4 and 5) was therefore the one for which the Contractors were invited to tender and the scheme which Rendel Palmer & Tritton have developed and finally designed, as described below.

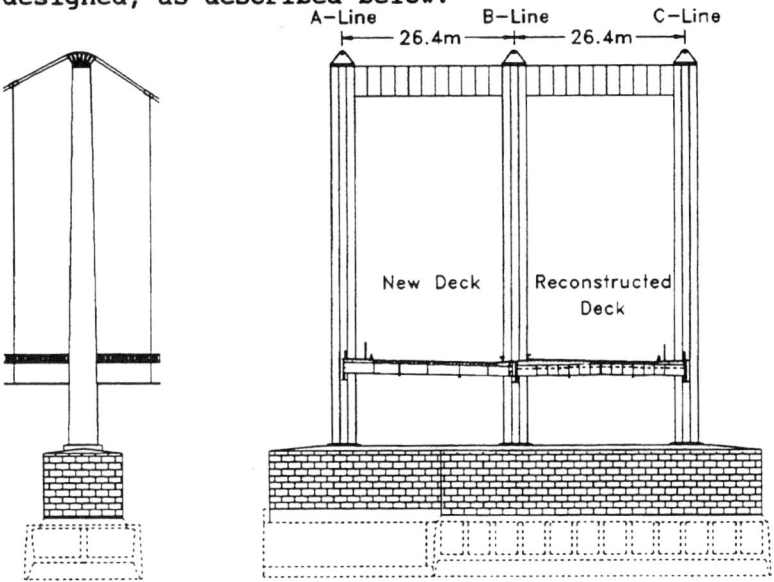

Fig.4 Widened and reconstructed tower elevations.

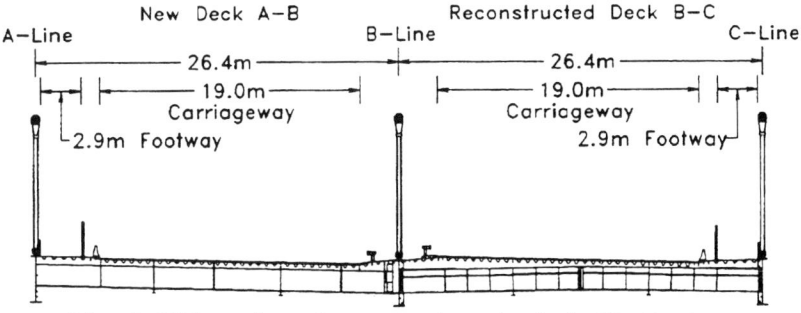

Fig.5 Widened and reconstructed deck cross section.

POSSIBLE REPLACEMENT OF EXISTING CABLES.

14. The chosen scheme will result in a final bridge in which a substantially new deck structure is supported by one new cable and two existing ones

that are over 40 years old. Initial investigations have shown that the condition of these cables is adequate. Nevertheless, the Federal Ministry of Transport required that, as a condition of the scheme being adopted, the feasibility must be established of replacing these cables at some future date without closure of the bridge. The basic methods by which this replacement of main cables would be carried out have therefore been established and, where necessary, additional strength has been incorporated into the design of the widened and reconstructed bridge.

15. It is proposed that the central cable of the bridge would be replaced first. This operation would be carried out with traffic restricted to two narrow lanes in each direction near the deck edges. The cable would then be simply removed and replaced. This condition, without a central cable forms the governing load case for the joint between the new cross girders and the existing bridge, the new cable saddles and anchorage structures amongst other items.

16. For the replacement of the existing outer cable, some of the ropes recovered from the central cable would be utilised as an auxilliary cable to provide temporary support to the deck during the operation. Although use of the bridge would be restricted, two narrow traffic lanes would be maintained throughout the cable replacement work.

FINAL DESIGN

17. The final design work has been carried out in accordance with German (DIN) standards current at the time of tender, with some minor amendments. These codes are base on allowable stress principles, limit state versions having only been adopted subsequent to the tender.

Loading

18. DIN 1072 specifies the loading on highway bridges, including the effects of wind and temperature. The client's technical specification adopts DIN 1072 loading except for some modifications which produce a slight reduction in the traffic loading for long loaded lengths.

New Main Cable and Hangers

19. The governing loading for the new main cable occurs during the proposed removal of the middle cable for replacement. To resist this load a slightly smaller area is required than that of the existing cables. However, the new cable has the same area as the existing cables so as to provide comparable stiffness for the same cable geometry. The new cable comprises 37 No. 69.83mm Dia. locked coil ropes rather than the 61 No. smaller ropes of the existing cables. This should result in greater durability.

20. The new hangers are 54.1mm diameter locked coil ropes and, as for the main cable, the governing load case is the loading with the middle cable removed. The arrangement of the new main cable and hangers is shown in Fig.6.

ROPE ARRANGEMENT – MAIN CABLE

Rope Diameter = 69.83mm
37 No. Locked Coil Ropes.

Cable Band

Hanger Clamp

Hanger
54.1mm Diameter
Locked Coil Rope

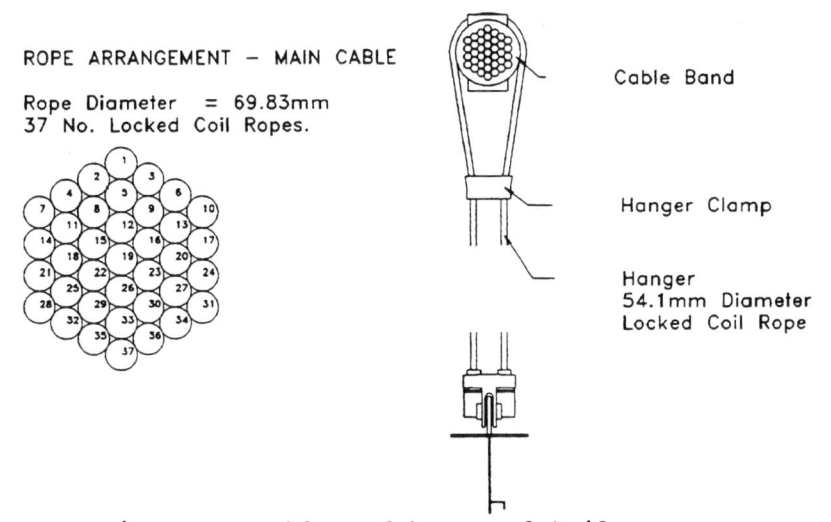

Fig.6 New cable and hanger detail.

Stiffening Girder

21. The stiffening girder is a welded plate girder with both horizontal and vertical stiffeners. The orthotropic deck plate is lap-welded to the inner edge of the girder top flange.

22. An economical design of stiffening girder is achieved by prestressing, which adjusts the stresses in the girder for the dead load condition so that the bottom flange size required is optimised. The required prestress is achieved by selecting a unstressed profile of the girder such that, when it is deflected into its final shape by the cable, the required stresses are developed.

New Deck Plate

23. The deck plate for the reconstructed and widened bridge comprises 12mm thick plate and 6mm thick trough stiffeners 225mm deep supported on the cross girders at 2.625m centres, which spacing matches the existing bridge cross girders.

Deck Cross Girders

24. The deck cross girders are welded plate girders which are simply supported between the stiffening girders during erection. This is achieved by providing a temporary articulation joint at the connection to the central girder. These articulation joints on each cross girder also accommodate the large relative movements of the cables during the widening and reconstruction. The cross girders are made continuous through the middle girder in the final condition.

25. The cross girder design is largely governed by the middle cable replacement case when the cross girders are required to span the full distance between the outer cables, some 52.8m. Surprisingly, this is only marginally more critical than the normal loading condition.

26. The articulation joint in the cross girder takes two forms (shown below in Fig. 7) in addition to the final permanent bolted connection. The pin connection allows the cross girder and deck plate to be erected. The deck plate is then welded to the stiffening girder flange as the pin is removed and replaced by a rubbing plate assembly. This allows the deck plate to act compositely with the stiffening girder while allowing the two decks to articulate.

27. Upon completion of the reconstruction of Deck B-C the connection is fixed by adding a bolted splice to the web and bottom flange; this allows load redistribution between the cables through the cross girders.

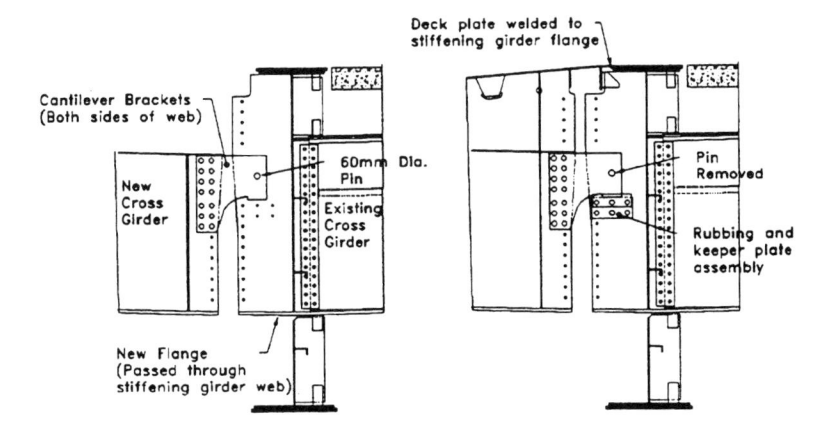

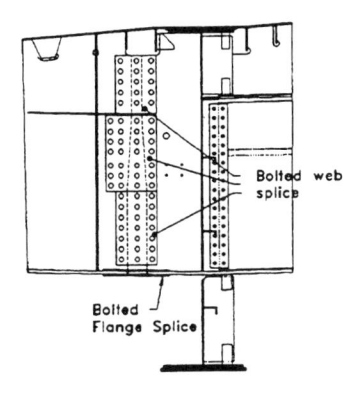

Fig.7 Articulation joint and permanent connection of cross girders to the B-Line.

Towers

28. The new tower leg and portal beam are of stiffened welded box construction and are connected to the existing tower via a hinged connection at the end of the portal beam. The hinged connection is required to limit the effects of differential settlement between the existing and the new foundations.

PART 1: WIDENING SMALL AND LARGE

Cable Anchor Frame

29. The cable anchor frame transmits the load from the main cable into the anchorage. The frame arrangement, shown in Fig.8 below, is such that the whole frame can be jacked along the axis of the cable by a substantial distance to simultaneously adjust the length of all ropes of the main new cable for the pull-in operation.

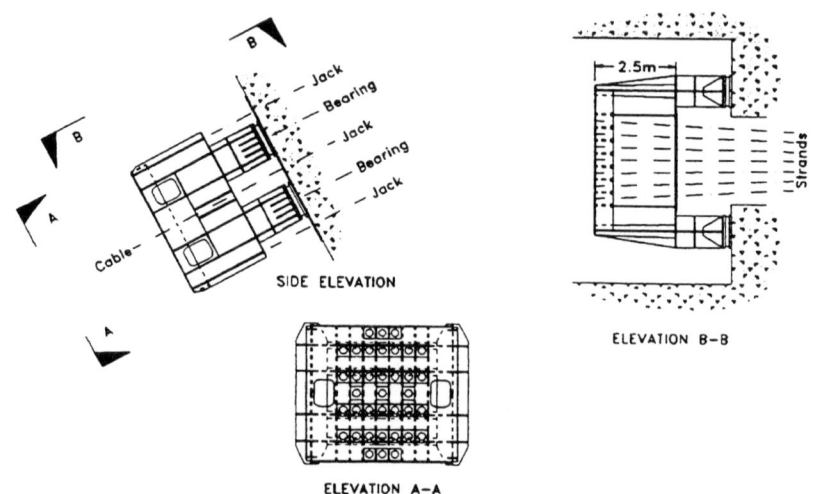

Fig.8 New cable anchor frame.

CABLE LOAD EQUALISATION

Requirement for Load Equalisation.

30. The reconstructed and widened bridge has been designed so that in its final condition the loads carried by each of the three cables will be approximately equal. This is a fundamental requirement requirement for the scheme to be workable for the following reasons:

31. *Overloading of the B-Line Cable.* During construction, the new and existing decks have to be hinged so that they can twist relative to one another about the central B-Line girder. This is necessary in order to isolate the traffic-carrying deck from the distortions due to the erection of the new deck. In the absence of such a hinge, erection of the new deck would impose unacceptable crossfalls on the traffic-carrying deck. In addition, variations in profile caused by traffic loading on the existing bridge would make the profile of the new structure difficult to control.

With this hinge, the new and existing decks are simply supported between each cable plane.

32. If the cable loads are not equalised, the existing middle, B-Line, cable would carry half of the total deck weight and the existing outer C-Line and the new A-Line cables would each carry a quarter of the total deck weight, as shown in Fig.9.

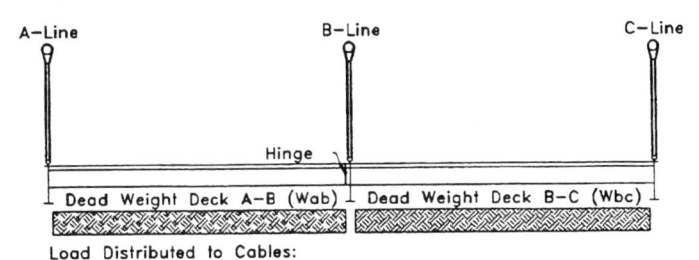

Fig.9 Distribution of dead load to cables Simple support.

33. As well as increasing the dead loading on the B-Line cable to an unacceptable level, the increase in dead load tension would also increase the vertical stiffness of the B-Line cable. This would then increase the proportion of live load carried by the B-Line.

34. <u>Transverse Profile of the Bridge.</u> If the bridge were built without cable load equalisation, the existing B-Line and C-Line cables would not have similar profiles as a result of the different loads that they would carry. In order to accommodate this misalignment in profile so that an acceptable carriageway crossfall is achieved, the deck plate would have to be supported by web upstands from the existing cross girders. Given that under these loadings the C-Line would be more than 1 metre higher than the B-Line at midspan, the height of these web upstands would become unacceptably large.

Achievement Of Equal Loads In Each Cable

35. <u>Analysis and Design.</u> The target is to achieve equal loads in each of the cables in the finished bridge. This is not in fact possible as the centre of gravity of the bridge deck is not exactly aligned with the centre of the bridge. This is primarily because the new stiffening girder is lighter than the existing ones. Due to this, adjustment of the 'target' loads carried by the cables is necessary. The adjusted 'target' loads

PART 1: WIDENING SMALL AND LARGE

are shown in Fig.10. It is important that the (unadjustable) existing B and C cables are equally loaded so that their final profiles are identical. Transverse equilibrium then determines the load that is carried by the (adjustable) new A-Line cable. Deck and cable profiles are derived for all three cables under these 'target' loads.

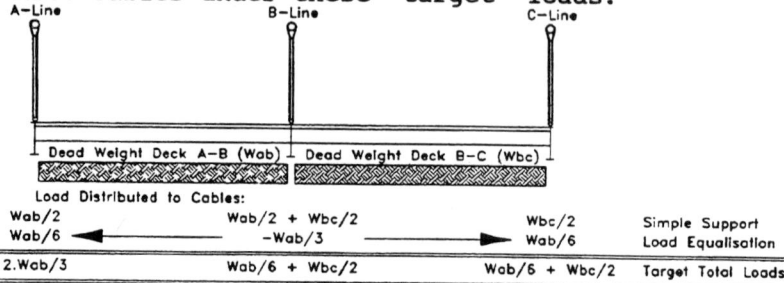

Fig.10 Distribution of dead load to cables Loads equalised.

36. Cable load equalisation is achieved by a pull-in of cable A after the erection of the new deck and the reconstruction of the existing deck have been completed. Up to this stage, the new cross girders are hinged at their attachments to the B-Line stiffening girder.

37. The loads carried by all three cables are determinate at this stage and thus profiles of the existing B-Line and C-Line cables and stiffening girders can be calculated. The required profile of the A-Line can then be determined by extrapolating the crossfall of the B-C deck to the A-Line as shown in Fig.11.

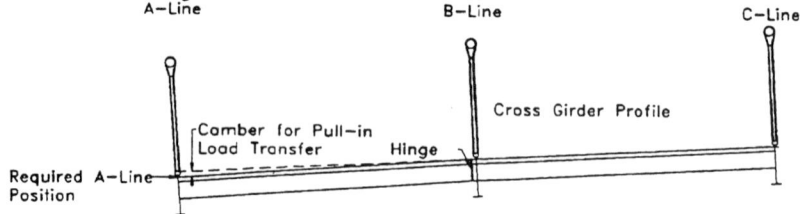

Fig.11 Required deck profile immediately prior to the pull-in.

38. The profile of the A-Line cable and anchor frame will be adjusted by movement of the cable anchor frame along the cable axis, thus effectively changing the cable length. To achieve the required A-Line profile, the anchor frames are 899mm in front of their final positions.

136

39. **Cable Pull-In.** The new deck A-B will be erected partially surfaced with the A-Line cable anchor frames 899mm in front of their final positions. At this stage, the connection between the new and the existing deck cross girders will be made rigid by replacing the 'hinge' by the final bolted moment connection shown in Fig.7.

40. Pulling-in of the A-Line cable will be achieved by jacking back by 899mm the cable anchor frames as shown in Fig.11. The cable load will be 25090kN at the start of the pull-in, increasing to 35700kN as the load equalisation occurs. The splay saddle simultaneously moves back 875mm on its sliding bearing. The difference between the movement of the frame and the splay saddle is accounted for by the extension of the ropes.

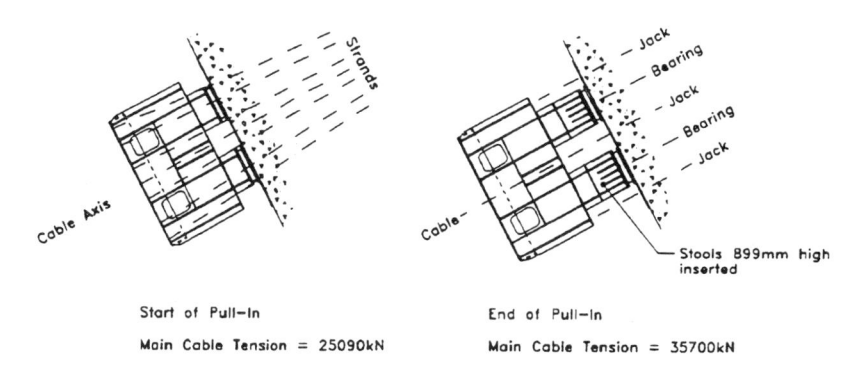

Fig.11 Cable anchor frame movement during pull-in.

41. At the start of the pull-in the tower saddles are positioned 750mm to the main span side of the tower centre. During the pull-in each saddle will be progressively jacked back relative to the tower, as shown on Fig.12, so that it will be concentric with the tower at the end of the pull-in. Movement of the tower saddles must be coordinated with those of the anchor frames to prevent tower overstress.

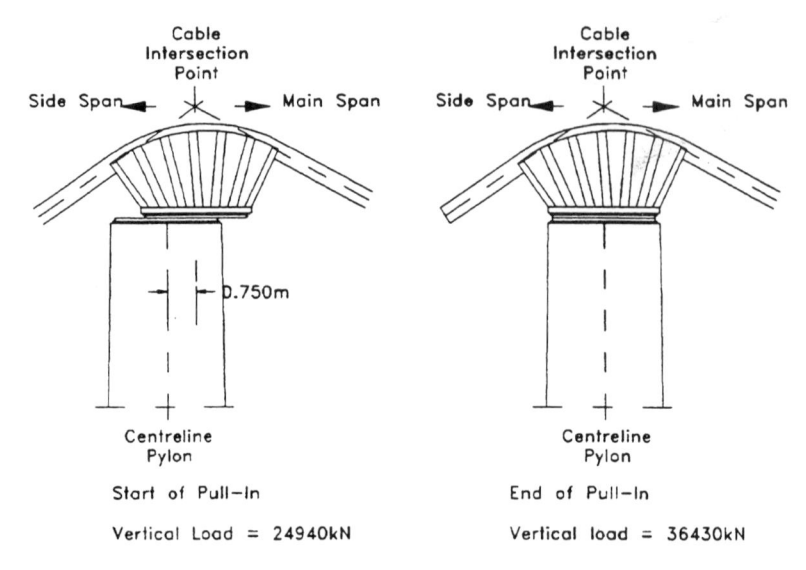

Fig.12 Tower saddle movement during pull-in.

SUMMARY
42. This paper has primarily covered the permanent works for the widening and reconstruction of the suspension bridge over the Rhein at Rodenkirchen. The presentation will concentrate on the erection engineering, the required staging of the construction to complete the bridge and a few of the more interesting design problems.

ACKNOWLEDGEMENTS
43. The original tender design for the widening and reconstruction was produced by the Engineering Department of the Landschaftsverband Rheinland. In January 1990, the design and construct contract was let by the Landschaftsverband Rheinland representing the Federal Ministry of Transport to a consortium consisting of Strabag Bau AG, Thyssen Engineering GmbH, and Cleveland Bridge & Engineering Ltd. Rendel Palmer and Tritton were awarded the design subcontract for the superstructure. Prof. Dr. Ing. E. h. K. Roik, Ingenieur-büro HRA is the Proof Engineer for the superstructure and Dipl. Ing. W. Jeromin, Köln is the Proof Engineer for the substructure.

15 year programme and requirements

P. H. DAWE, Head of Bridges Engineering Division, Department of Transport

SYNOPSIS The Department of Transport is undertaking a 15 year rehabilitation programme to bring its stock of trunk road bridges up to current standards. The programme which is dealing with various problems affecting bridges and their components is controlled through a comprehensive set of documents which outline the various technical requirements and organisational procedures.

INTRODUCTION

1. During the 1970's and earlier it was becoming clear that the Department's stock of bridges was deteriorating at a faster rate than had been anticipated. The publication of the new Bridge Assessment Code in 1984 had implications for the load carrying capacity of the short span bridges and indicated that some might not be adequate to carry current traffic with satisfactory margins of safety. Work on long span loading had also shown that the loads used to design these bridges were a lot lower than the loads possible from the greatly increased numbers of heavy goods vehicles now using the roads. In addition it was known that some structures had features which were not up to current standards and which could adversely affect the durability and safety of the structures. In view of the number of different and varied problems affecting the bridge stock it was decided that a concentrated effort should be made to tackle them. Accordingly a comprehensive programme of bridge rehabilitation for motorway and other all purpose trunk roads was developed, which was announced by the then Minister for Roads and Traffic, Peter Bottomley, in November 1987.

2. In developing the programme three strategies for completion were investigated, with the work spread over 10, 15 and 20 years, together with the corresponding estimated costings. In respect of the 10 year strategy it was found that high peak levels of spending were required and it was felt doubtful whether these could be achieved in practice. The spend profiles for 15 and 20 years were not significantly different, but it was considered that because of the safety implications and the need to get our structures ready for the

PART 2: POLICY

heavier EC vehicles by 1999 the shorter 15 year period was preferable. The various items of work in the programme may be aggregated into the following general groups:-

   (i)   Assessment and Strengthening
   (ii) Steady State Maintenance (deterioration)
   (iii) Upgrading of sub-standard features

The paper will describe the various items of work and the documents which have been produced which set out the technical requirements for the work.

ASSESSMENT AND STRENGTHENING
   3. This part of the programme is concerned with upgrading the national bridge stock to meet the loading requirements derived from current and proposed vehicles. Bridges are being assessed and strengthened to carry lorries up to the EC weight limits of 40 tonnes gross and 11.5 tonne for drive axles to meet the end of the UK derogation on lorry weights on 31 December 1998. The necessary assessment and strengthening work will be carried out in three, largely concurrent, stages. The assessment criteria also cover 2, 3 and 4 axled vehicles whose weights are affected by the EC directives and which are also covered by the derogation.
   4. <u>Stage 1 - Older Short Span Bridges.</u> This stage is for the pre-1922 short span bridges and post-1922 short span bridges not known to have been checked for 30 units of HB. The 1922 date was the date when a national highway loading was first introduced. Because these bridges are the ones most affected by the move to the heavier EC vehicle weights this stage has been given priority with the assessments to be completed within the first three years and the strengthening to be completed by 1996. The Bridge Assessment Code, Departmental Standard BD 21/93 and Advice Note BA 16/93, form the basic reference documents for this work, but the technical requirements for the particular stage are given in BD 34/90 and BA 34/90.

   [Note: References to all Departmental Standards and Advice Notes quoted in this paper are given in Annexes 1 and 2.]

   5. <u>Stage 2 - Modern Short Span Bridges.</u> This stage covers some more modern bridges which were designed before the enhanced shear requirements for reinforced concrete and prestressed concrete slabs and beams were introduced in 1973. The technical requirements for this stage are given in BD 46/92. The assessment criteria for concrete bridges are given in the assessment version of the national bridge code, BS 5400: Part 4, namely BD 44/90 and the complementary Advice Note BA 44/90. This assessment version contains such relaxations as are felt possible for calculating the shear strength of concrete elements.

6. Stage 3 - Long Span Bridges. Long span bridges includes bridges with loaded lengths greater than 50m which are affected mainly by the large increases in the numbers of heavy goods vehicles rather than by increases in vehicle weights. The technical requirements for this stage are given in BD 50/92 which refers to the Department's loading standard BD 37/88. However in cases where there are low traffic flows or low percentages of heavy goods vehicles, assessment loading criteria can be developed for the particular bridge in question using the relevant site traffic data.

STEADY STATE MAINTENANCE

7. The core of this area of work is the continuing programme for the repair and replacement of the various parts of structures which have deteriorated with time and use. It covers many items which require regular treatment and which will continue to need attention even when the 15 year programme is completed. In particular the steady state maintenance section aims to implement many of the recommendations of the Maunsell report (ref. 1) on the state of the Department's concrete bridges, and includes both preventative maintenance activities as well as remedial work. The area of work includes the repair of deteriorated concrete, painting of steel structures, the routine replacement of bearings and joints, and the replacement and repair of waterproofing systems.

8. Preventative maintenance includes the surface impregnation of concrete elements and the repair and replacement of leaking expansion joints. This work is being given priority because both activities have been identified as two of the most effective ways of reducing further chloride penetration. The inspection and repair of concrete structures is dealt with in BA 35/90 while the specification for the materials for concrete repair is given in BD 27/86. The criteria for the impregnation of concrete on both new and existing structures are given in BD 43/90 and BA 33/90. It is recognised that cathodic protection is a viable and effective method in certain circumstances for dealing with structures which contain a high level of chlorides. However the result of trials now underway need to be fully assessed before any formal advice on its wider use is issued.

9. Alkali silica reaction (ASR) has been recognised as a possible cause of deterioration which affects bridges mainly in two areas of the country. Recent research work seems to indicate that the problem is one more of durability rather than loss of strength and that what is needed is a management strategy for those found to be affected. Advice on the diagnosis of ASR has been issued by the British Cement Association (ref. 2) and the Institution of Structural Engineers (ref. 3) have issued interim guidance on the appraisal of ASR affected structures. Procedures for the management of ASR affected structures are given in BA 35/90.

PART 2: POLICY

10. The Maunsell report also recommended increased levels of investigation for structures. This proposal has been met by incorporating material testing into the normal Principal Inspection programme with the requirements being given in BA 35/90. It was expected that this increased level of inspection would initially be for a complete round of Principal Inspections i.e. for about six years.

UPGRADING OF SUB-STANDARD FEATURES

11. This part of the programme is intended to rectify deficiencies in certain structures where current design standards or specifications are not met, mainly those affecting safety and/or durability. It also deals with certain problems which have been identified in prestressed concrete bridges. The following topics are intended to be dealt with under this heading:-

- (i) Waterproofing unprotected bridge decks

- (ii) Rehabilitation of post-tensioned concrete bridges

- (iii) Repairs to deflected tendon PSC bridges

- (ix) Replacement of sub-standard parapets

- (v) Counter measures (structural) to 'bashing' of low headroom bridges

- (vi) Strengthening of piers and columns to resist higher impact forces

- (vii) Health and Safety aspects of access to bridges.

12. <u>Waterproofing unprotected bridge decks.</u> The Maunsell report identified a good waterproof membrane as a vital defence against contamination by salt. Although waterproofing of bridge decks has been mandatory since 1967 it is known that there are a number of bridges built before then which have not been waterproofed or where waterproofing is only partial eg only under verges and footways. Waterproofing is being given priority under the programme in order to prevent further chloride penetration and further deterioration.

13. <u>Rehabilitation of post-tensioned bridges.</u> Post-tensioned bridges were originally included in the 15 year programme as a possible item for future work pending further studies. Some evidence was available to suggest that a number of tendon ducts had not been properly grouted but it was not known how significant this might be for the durability and safety of the structures. However since then a number of bridges have been found with severe corrosion of the prestressing tendons and it has been decided to carry out a programme of special inspections of all post-tensioned

concrete bridges with bonded tendons. This programme which will be carried out over 5 years was announced by the Minister for Roads and Traffic, Kenneth Carlisle in September 1992. To aid in the implementation of this programme a number of documents have been (or will shortly be) issued. BD 54/93 gives the criteria for the prioritisation of the special inspections and BA 50/93 gives advice on the planning and organisation of the special inspections and describes the various non-destructive methods which are available. A further advice note on ways of repairing and strengthening post-tensioned bridges is being drafted. In addition to the various documents the Transport Research Laboratory, in conjunction with Bridges Engineering division has also held a number of seminars about the inspection of post-tensioned structures. These have included demonstrations of some of the more promising non-destructive inspection techniques.

14. Repairs to deflected tendon PSC bridges. It was discovered that the cement mortar filling to pockets situated at the deflector points in the soffits of some pre-cast beams contained calcium chloride added as an accelerator The chloride had caused corrosion of the deflectors and prestressing strands and spalling of the concrete. As it was not a widespread problem, and because the bridges affected could be easily identified, it was not thought necessary to publish any general instructions for dealing with the problem. However trial repairs were carried out in an attempt to identify the most effective method of repair.

15. Replacement of parapets. Some action was taken in advance of the 15 year programme to replace sub-standard bridge parapets at particularly high risk sites. Since then all sub-standard parapets have been included in the programme and are being replaced in accordance with the prioritisation criteria set out in BA 37/92. The advice note also gives guidance on alternative methods of protection using safety fences.

16. Bashing of bridge decks by high vehicles. Bashing of low headroom bridges has been examined by a Departmental Working Party which published its report in 1988 (ref 4). The report suggested a strategy for substantially reducing the incidents over 5 years. Most of the attention is focused on the improvement of signing at low bridges, the installation of automatic over-height warning devices and legislation requiring vehicle heights to be fixed in the driving cab. However a departmental standard is being developed with assistance from British Rail for the design of bridge protection beams. These beams are erected just prior to any low headroom soffit but are attached to the bridge's own foundations.

17. Strengthening of piers and columns. Many bridges have supports which were designed for fairly nominal impact forces and are no longer adequate to deal with the collision loads from current heavy goods vehicles. All existing bridge supports are being assessed for enhanced collision loading

PART 2: POLICY

criteria which are also being applied to new design. The requirements for the assessment and strengthening of bridge supports are given in BD 48/93, which includes provision for adopting protective measures rather than strengthening the support.

18. <u>Health and Safety aspects of access to bridges.</u> A number of box girder bridges have been identified where the size and/or position of the access manholes do not satisfy current HSE requirements. These are being modified but there are other aspects of access to highway structures which need to be reviewed in the light of current HSE requirements. An advice note which deals with this topic is currently under preparation. This work is not being centrally co-ordinated but left to the regions to incorporate as they see fit in their annual programmes.

19. <u>Programming.</u> The actual programming and funding of the work is the responsibility of NGAM division operating through the regional NMD's. Instructions to the Departments agent are given through Trunk Road Management and Maintenance Notices and the implementations of the 15 year programme was dealt with in TRMM 2/90. This points out the importance of co-ordinating the different remedial work activities which have been found necessary for a particular structure and also with co-ordinating where possible bridge activities with other planned highway maintenance to minimise traffic disruption. The programming of the work will also be affected by the Motorway Widening programme where it is expected that about 1000 bridges will need to be replaced. Obviously some of these bridges will also be those earmarked for remedial work under the 15 year programme.

STRENGTH ASSESSMENT

20. As originally written the Bridge Assessment Code referred to the relevant parts of the national bridge design code BS 5400 for assessing the resistance/strength capacity of the various structural elements. Although this is a modern code and incorporates the findings of up-to-date research there are certain disadvantages in using a design code for assessment purposes:-

(i) Since a degree of conservatism is not a great disadvantage in the design of new works, there is a tendency for design code drafting committees to make fairly conservative assumptions, especially where there may be uncertainty about the interpretation of research data or about its translation into design rules. This also applies when the rules may be used for a wide variety of structural types, not all of which the drafters may be able to foresee at the drafting stage. Values can also be rounded off to make presentation neater or lower bounding curves fitted to fairly scattered experimental results.

These conservative simplifications, made at different stages of the calculations, may be multiplied together, and thereby augmented considerably as the design proceeds. However, a conservative solution which would add only a small percentage to the cost of a new bridge as designed, may, in an assessment, mean the difference between concluding that the structures is adequate, or capable of strengthening and the conclusion that the bridge must be replaced with very large cost implications.

(ii) A design code deals with undesirable structural types or details by simply prohibiting their use. This is not, of course, appropriate in assessment because the assessing engineer must deal with the structure as it exist. Also as detailing practices change so the earlier details become out-of-date and are no longer covered by the codes.

(iii) Whereas a design code, in setting values for the various partial factors, has to make allowance for variations in material properties, section sizes etc of a structure which does not yet exist, the assessing engineering has the facility of being able to measure these properties in the actual structure. Of course, there are limits to the extent to which this can be done on existing structures, but nonetheless, some relaxation in partial factors can be made, depending on the degree of certainty that the assessing engineer is able to reach about these material values.

21. It was therefore decided to develop assessment versions of the relevant parts of the design code in an attempt to eliminate as much as possible of the built-in conservatism while still adhering to the basic target level of reliability for the bridge stock as a whole. The three parts of BS 5400 identified as being in need of the treatment were Part 3 (steel), Part 4 (concrete) and Part 5 (composite), with Part 4 being the one most urgently needed. Accordingly an assessment version of Part 4 was commissioned and was published as a departmental standard and advice note in 1990 in BD 44/90 and BA 44/90 respectively. Assessment versions of Part 3 and Part 5 have also been undertaken and will be published shortly. It should be noted that there is now no independent assessment live loading since it is derived directly from the design loading by the use of appropriate factors to give the standard assessment loading and the reduced levels of loading. The design loading

PART 2: POLICY

contains a 10% contingency margin to allow for future changes in traffic patterns and this is removed when assessing existing structures.

RESEARCH

22. The rehabilitation of an existing stock of bridges reveals how little we know about the behaviour of full-size structures and components and about the factors which affect their durability. Research therefore has a very important part to play in providing answers to some of the problems to ensure that any remedial work is carried out in any effective and economic manner. The development of the various documents described in this paper has been supported by an extensive programme of research covering all aspects of assessment and repair.

23. The programme has included the load testing of full-size redundant structures and elements, such as beams, as well as the investigation of effective methods of repair and repair materials. The research has also sought to improve some of the effectiveness of materials and components which have a direct effect on durability. For instance a comprehensive programme of work has been undertaken to develop better criteria for assessing the performance of bridge deck waterproofing systems. Machines have been developed which will allow joints to be tested under both translational movements and wheel loads. In both cases the use of more effective and longer lasting systems will significantly increase the durability of the parent bridge.

24. The findings of the current research will be used to amend the standards and advice notes as and when the results become available. For instance there is a project investigating the impact forces developed when vehicles collide with bridge supports. This involves not only computer modelling but also full-scale impact tests to provide information to calibrate the models. The result of this work should provide a sounder basis for assessing bridge supports for their resistance to impact and will also be relevant for new design.

CONCLUSION

25. Bridges are long life structures which must be expected to be subjected at some time during their service lives to loads and other demands which were not foreseen at the time of their design. Sometimes there are sufficient reserves of strength and durability built into the structure to be able to cope with these unexpected demands but sometimes there are not. Technical standards may change following new research or public expectations regarding safety may alter, all of which may mean that some existing structures don't fully comply with the current requirements. Whatever the cause of the change it may not be possible to deal with them in the regular maintenance programme and it will be necessary to mount a special exercise to deal with

the shortfalls and bring the stock of bridges up to current standards within a reasonable time.

26. This paper has described how such an exercise is being carried out for the bridges, and other structures, on the UK trunk road system. Because of the fairly complex organisational structure through which the work is carried out and the number of different organisations involved it has been necessary to develop a comprehensive set of procedural and technical documents to direct and control the way that the work is carried out. In many ways the organisational and procedural documents are just as important as the technical criteria for ensuring that the work is carried out in an orderly, cost effective and consistent manner.

27. Despite all the guidance and technical advice that can be issued it must be recognised that the management of an inhomogeneous stock of highway structures can be more of and art than an exact science and does require the engineer to exercise a great deal of judgement. This is particularly true for older structures which use unfamiliar forms of construction and where detailed aspects of the design are not covered by current codes. Thus there is often greater uncertainty about dealing with existing structures than there is with the design of new structures which means that the mainly deterministic and prescribed approach adopted for new design is not always appropriate. There is an example of this in the assessment version of the concrete code where the assessing engineer is allowed to estimate a worst credible strength for the materials for use in the calculations, rather than accept a value given in the code.

28. However the development of the reliability methods and techniques means that there are now available powerful tools for dealing with uncertainty especially in the structural field. These have been used to a limited extent for determining the reliabilities of existing structures which have deteriorated in some way or other, but there is some way to go before the reliability approach can be adopted as a technique for general application by practising engineers. Although it would be unwise to rely on the absolute values generated by any reliability calculation the method does provide a rational way for comparing the risk involved in different situations. Thus the reliability of a deteriorated concrete bridge could be compared with one in pristine condition and the assessment engineer given a better basis for exercising his judgement about the remedial measures which might be necessary. Reliability methods are also able to tackle a wide range of problems where risk or uncertainty is involved and are thus very suitable for the highway situation where structures and components are faced not only with carrying varying stochastic loads but also have to resist the chance collision from errant vehicles. It is hoped that the limited use of the reliability method made so far in this country for calibrating design codes can be extended to the bridge management field, so that this difficult art can become more of a science.

PART 2: POLICY

REFERENCES

1. WALLBANK E. J. (G Maunsell & Partners). The performance of concrete in bridges. HMSO, London, 1989.

2. The diagnosis of alkali-silica reaction. Working party report.
British Cement Association, Slough, 1988

3. Structural effects of alkali-silica reaction. Working party report.
Institution of Structural Engineers, London, 1992

4. A Strategy for the reduction of bridge bashing. Working party report.
HMSO, London, 1988

ANNEX 1

Documents published by the Department of Transport in support of the 15 year rehabilition programme

General

Trunk Road Management and Maintenance Notice TRMM 2/90. All purpose trunk road and motorway structure maintenance and inspection expenditure, 5 year rolling programme for major schemes; 15 year rehabilitation programme.
April 1990.

Assessment and Strengthening

Departmental Standard BD 21/93. The assessment of highway bridges and structures. January 1993.

Advice Note BA 16/93 The assessment of highway bridges and structures January 1993.

Departmental Standard BD 34/90 Technical requirements for the assessment and strengthening programme for highway structures. Stage 1 - Older short span bridges and retaining structures. September 1990.

Advice Note BA 34/90 Technical requirements for the assessment and strengthening programme for highway structures. Stage 1 - Older short span briges and retaining structures. September 1990.

Departmental Standard BD 46/92 Technical requirements for the assessment and strengthening programme for highway structues. Stage 2 - Modern short span bridges. August 1992.

Departmental Standard BD 50/92 Technical requirements for the assessment and strengthening programme for highway structures. Stage 3 - Long span bridges. December 1992.

Departmental Standard BD 44/90 The assessment of concrete highway bridges and structures. October 1990.

Advice Note BA 44/90 The use of Departmental Standard BD 44/90 for the assessment of concrete highway bridges and structures. October 1990.

Advice Note BA 38/93 Assessment of the fatigue life of corroded or damaged reinforcing bars. April 1993.

Advice Note BA 39/93 Assessment of reinforced concrete half-joints. April 1993.

PART 2: POLICY

Steady State Maintenance

<u>Departmental Standard BD 27/86</u> Materials for the repair of concrete highway structures. November 1986.

<u>Advice Note BA 35/90</u> The inspection and repair of concrete highway structures. June 1990.

<u>Departmental Standard BD 43/90</u> Criteria and material for the impregnation of highway structures. April 1990.

<u>Advice Note BA 33/90</u> Impregnation of concrete highway structures. April 1992.

Upgrading Sub-Standard features

<u>Advice Note BA 37/92</u> Priority ranking of existing parapets, October 1992.

<u>Departmental Standard BD 48/93</u> The assessment and strengthening of highway bridge supports. June 1993.

<u>Departmental Standard BD 54/93</u> Post-tensioned concrete bridges. Prioritisation of special inspections. January 1993

<u>Advice Note BA 50/93</u> Post-tensioned concrete bridges. Planning organisation and methods for carrying out special inspections. July 1993.

DAWE

ANNEX 2

Documents to be published by the Department of Transport in support of the 15 year bridge rehabilitation programme

Subject	DS	AN	Publication
Assessment and Strengthening			
Assessment Version of BS 5400: Part 3: Steel	✓	✓	Spring '94
Assessment Version BS 5400: Part 5: Composite	✓	✓	Early '95
Assessment of structures affected by Salt and ASR		✓	Spring '94
Assessment of Sub-structures and foundations		✓	Early '94
Load Testing for Assessment		✓	Early '94
Steady State Maintenance			
Strengthening with bonded plates		✓	Early '94
Cathodic Protection		✓	1995
Sub-Standard Features			
Repair and strengthening of post-tensioned structures		✓	Spring '94
Access and security of bridges		✓	Autumn '94
Bridge protection beam	✓		Spring '94

DS = Departmental Standard
AN = Advice Note

The management of the assessment and strengthening programme

A. LEADBEATER, Deputy County Engineer, Oxfordshire County Council

SYNOPSIS. This paper describes the management of the 15 year rehabilitation programme announced in 1987. Local Authorities (County Councils, Regional Authorities and Metropolitan Districts) have responsibility for managing the programme as it affects all their own bridges, with the Department of Transport, the Scottish and Welsh offices for the majority of trunk road and motorway bridges and with the British Waterways Board, British Rail and Scotrail programming the work to ensure the least effect on customers who use the road system in the UK. Comparisons with the late 1960's bridgeguard programme will be made, progress and problems encountered will be described and issues arising from the programme will be discussed. The conclusion is that it is vital that Government support is given to the programme and that Local Authorities must ensure that the opportunity is fully grasped to ensure that bridge (and retaining wall) stock is in the best possible condition when the programme concludes.

DOCUMENTATION

1. The Department of Transport has issued two documents relevant to the programme TRMM 2/90 Appendix 2 (concerning motorways and trunk roads) and Circular Roads 2/91 (concerning Local Authority roads plus private bridges carrying Local Authority Roads eg. BR, Scotrail, BWB and other private bridges). Key dates given in those documents are as follows:

Government bridges will be assessed and strengthened to carry lorries up to the EC limits of 40 tonnes (gross) and 11.5 tonne drive axle by 31 December 1998.

Other elements on Government bridges such as repainting, waterproofing, concrete repair and upgrading of post tensioned bridges etc, will be completed by 2004.

2. No programme is provided for Local Authority and private bridges but clearly it would be impractical and folly to restrict the 40 tonne construction and use vehicles to

trunks roads and motorways only - virtually all journeys start and finish on Local Authority roads. The conclusion clearly is that Local Authorities must ensure that their programme if not co-incident with the Government's programme follows closely behind.

BRIDGEGUARD PROGRAMME (Late 1960's)

The Bridgeguard exercise commenced in the late 1960's was the first major, national bridge refurbishment exercise and task. It arose not due to an increase in long and/or axle weight but due to the unexpected collapse of structures affecting the nation at large (not, however, highway bridges). The strengthening work was supported by the Government through the grant system them in operation. Unfortunately, strengthening was patching and very many bridges were not dealt with. This was because at that time, very many Local Authorities were responsible for bridges eg. in one large authority 38 different LAs had Highway Authority responsibility - replaced, of course, by 1 in 1974. (It is interesting to surmise whether the much smaller Unitary Authorities proposed in the Local Government Review may have the same problems!) The differences are summarised below:

Bridgeguard (1967 - 1974)	15 year Rehabilitation Programme (1987 - 2002)
Simple Assessment Code BE4 (16T or 24T vehicles)	Complex Assessment Code (BD 21/84, BA 16/84)
Simple analytical methods (Hendry & Jaegar, Maurice & Little etc)	Complex analytical methods (Computer analysis etc)
Little knowledge of material breakdown (Corrosion & rusting of steel)	Great knowledge of material breakdown (ASR, Chloride attack etc)
Few bridges traffic sensitive	Many bridges traffic sensitive
Supportive public	Questionning public
BR and BWB responsible for full strengthening	BR and BWB responsible for strengthening to BE4 only

PRIVATE BRIDGES

4. The following category of private bridge types carry Local Authority roads:

i) British Rail (& Scotrail) - can include disused lines
ii) British Waterways Board
iii) London Underground Limited
iv) Private bridges owned by individuals, private companies, operated as leisure routes (eg. National Park or Local Authority owned) and private railways.

5. Category i) and iii) bridges above are required to strengthen structures they own to BE4 (1968) standards only.

PART 2: POLICY

The cost of strengthening a weak bridge to present day standards from BE4 standards lies with the Local Authority. To all intents and purposes category ii) bridges can be considered the same. No guidance is given on Category iv) bridges. Presumably liability for assessment/weight restriction if necessary/strengthening lies with the Local Authority. Who pays will be dependant on local agreement.

6. A sample of 59 Authorities in England, Scotland and Wales (excluding London) showed that the following numbers of "private bridges" exist in the various categories carrying highways:

Category 1 - 3067
Category 2 - 875
Category 4 - 2013   (Category 3 - no return from London U'ground)

7. The ruling that Local Authorities are responsible for maintaining and paying for strength above BE4 (1968) levels could have extreme consequences for future expenditure.

8. It is also unclear as to how Highway Authorities proceed with Category 4 bridges owned by others (not statutory authorities). What is certain is that the Local Authority will need to ensure that these bridges too, meet the latest standards and criteria.

PLANNING THE PROGRAMME

9. The method of Planning the Programme for trunk roads and motorway bridge structures is clearly set out in TRMM 2/90 Appendix 2. This programme covers the complete rehabilitation of all bridges owned by the Government (numbering approximately 13,000). In conjunction with the pavement strengthening programme and bearing in mind that the majority of Government bridges are "modern" ie. post-war the programme is relatively easy to plan. The Principal Inspection system will also provide excellent background information and the majority of problem bridges may have been identified.

10. Turning to Local Authority Bridges (numbering approximately 129,000) the variety, age, type and condition of each bridge is almost infinite.

11. Private bridges (Rail and BWB) are generally metal or masonry/brick structures up to 200+ years old. Little work has commenced on rail bridges to date, a source of concern for those responsible for managing the Local Authority highway system (it should be recalled that under various Trunk Road Acts the Government took over responsibility for the majority of bridges carrying trunk roads over railways and canals).

12. It can be seen therefore that the problems associated with planning the Local Authority programme are extensive and complex.

BRIDGE DECK TYPES (Typical Midland County)

13. Stone Slabs 94   Steel Beam Jack Arch 22
    Insitu RC 282   Pre-stressed concrete 99
    Post-tensioned concrete 2   Pre-cast box 42
    Corrugated stell 46   Steel Beam/Troughing 12
    Arches 588 (of which 319 do <u>not</u> conform to NEXE, ie multi-span, widened, gothic, spine beam etc, etc.)

PRIORITISATION

14. A first step in planning the programme is to produce a priority system for the programme of assessment. A sample of the majority of Local Authorities in the UK reveals that County Bridge Engineers priority rating is based on bridge type with a nearly 60% choice of metal decks as first priority (see Fig.1). However, many Authorities have chosen other priority systems - the most popular being on a strategy basis ie. all bridges primary routes, then other "A" routes, then "B" routes etc, etc.

15. All County Bridge Engineers have, of course, dealt with known problem bridges first.

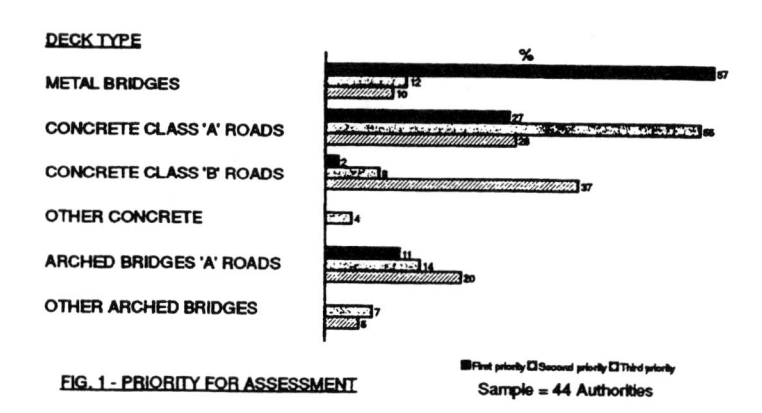

FIG. 1 - PRIORITY FOR ASSESSMENT

Sample = 44 Authorities

PART 2: POLICY

ASSESSMENT PROCESS

16. Individual bridge assessment can be simple or highly complex depending on how near the limits the structure may be and on condition factor of the materials which form the structure. A simple flow diagram is shown which demonstrates a very simple system. The two items I highlight are:

a) the top half of the process is analysis only whereas below that analysis is accompanied by engineering judgement.
b) the difficulty of assessing material condition on the final assessed strength of the structure.

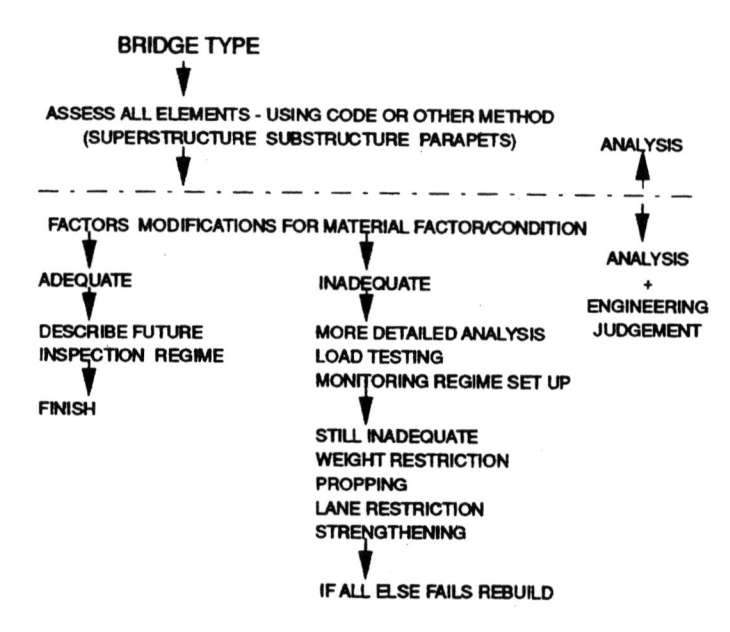

```
                        BRIDGE TYPE
                             ↓
          ASSESS ALL ELEMENTS - USING CODE OR OTHER METHOD
            (SUPERSTRUCTURE SUBSTRUCTURE PARAPETS)         ANALYSIS
                             ↓                                ↑
         — — — — — — — — — — — — — — — — — — — —             |
            FACTORS MODIFICATIONS FOR MATERIAL FACTOR/CONDITION
                             ↓                                ↓
                                                          ANALYSIS
         ADEQUATE              INADEQUATE                    +
            ↓                      ↓                    ENGINEERING
         DESCRIBE FUTURE      MORE DETAILED ANALYSIS     JUDGEMENT
         INSPECTION REGIME    LOAD TESTING
            ↓                 MONITORING REGIME SET UP
         FINISH                   ↓
                              STILL INADEQUATE
                              WEIGHT RESTRICTION
                              PROPPING
                              LANE RESTRICTION
                              STRENGTHENING
                                  ↓
                              IF ALL ELSE FAILS REBUILD
```

FIG. 2 - ASSESSMENT PROCESS

17. In conclusion it is vital that a well organised bridge management system is set up to ensure that the processes used and assumptions made in the assessment process are clearly available to future assessment engineers and that a proper regime of inspection etc is indicated.

PROBLEMS OF ASSESSMENT

18. It is quite clear that problems of interpretation of any code will inevitably occur - this will be magnified in the assessment process where almost everyone of the 160,000

highway bridges has some unique factor. This is not the place to discuss these issues in detail, but some of the principal problems identified in the County Surveyors Society survey are itemised below:

i) **Masonry Bridges**   Assessment of spandrels, problems of multi-span arches and widened arches, assessment of sub-structure, assessment of granite structures.

ii) **Reinforced Concrete**   Temperature effect on RC arches, sub-structure (including impact), making short span bridges "work", method of analysis grillage v simple, shear problems generally.

iii) **Composite Structures**   Effect of composite action of previous design or previous strengthening

iv) **General**   Modelling of structures, material factors, serviceability departure from standard, extent of testing and inspection.

19. The above list is by no means complete, but gives a flavour of the type of problem occurring. What is certain, in my view, is that a slavish adherance to the Codes will result in pessimistic assessments, increase in temporary or permanent weight restrictions and the waste of materials when potentially sound structures are demolished.

PROGRESS TO DATE

20. Again, the following information is based on the CSS survey as at 1 July 1993. This survey covers the majority of Counties and Metropolitan District Councils in mainland Britain:

21. i) DOT/SDD/WDD Structures

2301 bridges assessed, out of which 1407 require no further action. 43 have been rebuilt. 228 are being monitored. The latest anticipated complete date for assessment is given as 1997 and it is believed that the strengthening programme will be complete in the year 2000. Virtually all Local Authorities consider sufficient funding is being provided to meet the programme outlined in paragraph 1 above. Subjectively, it is reckoned that significant traffic problems have occurred at 40 sites due to the programme (please note the programme statistics exclude London Boroughs and London Underground Bridges). It should also be noted that the Government Departments have taken over all highway bridges over railways, canals and other private bridges so are not affected by problems with the Railway Companies.

PART 2: POLICY

22. ii) Local Authority Structures

11,959 bridges assessed so far, out of which 6,452 bridges require no further action, other statistics as below:

Number of bridges strengthened - 556
Number of bridges weight restricted - 348
Number of bridges rebuilt - 339
Number of bridges being monitored - 627
Number of bridges cleared using load testing - 19
Significant Traffic Problems have occurred at 129 sites

Number of BR/Scotrail bridges assessed - 78 out of 5067 (1.5%)

23. Local Authorities consider that the assessment programme will be complete for all their bridges in the range 1994 - 2003 years. Local Authorities consider that the strengthening programme will be complete for all their bridges in the range 1998-2005 years.

24. Those dates exlude the likely effect of the BR/Scotrail programme where insufficient data to make a projection for completion is available.

25. Whilst two-thirds of the Authorities surveyed were satisfied with the level of support from Government sources for their programmes, less than 20% gained their full bid.

26. The three main concerns over Local Authority bridge strengthening are therefore over finance, the extended programme and the effect of the British Rail programme (when it commences in earnest). Further delays may, of course, occur as a result of Local Government re-organisation.

COMPARISON WITH SAMPLE SURVEY RESULTS

27. In 1987 all the major highway bridge owners co-operated in a bridge census and sample survey to attempt to guage the scale of the problem (excluding Government bridges). The CSS survey mentioned in this paper is, I believe the first time that efforts have been made to see how the programme is progressing. The following indicate broad comparisons to date.

Census (1987)   CSS Survey (1993)

i) 11260/54171 = 21%   5457/11959 = 45%
Proportion of bridges
below standard (Table S1 & Table S3)

ii) £378 million   May be as much as double
Cost of strengthening LA bridges
(Table 20)

28. The above conclusion must be considered very approximate only. The reason for the apparent increase in sub-standard bridges may be as a result of several factors listed below:

a) more servere assessment code than anticipated (eg. shear failure in concrete).

b) more detailed system of assessment and greater appreciation of material and serviceability conditions·

c) I have included all the bridges being monitored as below standard - otherwise why do it? However, I believe many bridges being monitored will eventually be cleared with no further work.

d) Higher failure rate of steel bridges (0.66 in the survey) and concrete bridges (0.35 in the survey) than could be reasonably anticipated.

What is clear is that more strengthening and more expenditure is likely than anticipated in the sample survey.

OTHER ISSUES

29. There are several other issues which are very relevant to the programme and these are outlined below.

30. Retaining walls and cellars. Many kilometres of rural road are supported or protected by old masonry retaining walls (sample survey Appendix B shows at least 5,400 kilometres in length retaining over 1m of fill. The assessment of these walls is an "art" rather than "a science". It is right, in my view, however that the assessment and strengthening programme should include strengthening necessary to retain the highway network. At least 65% of Authorities surveyed felt that at least £100,000 needed to be spent on retaining walls as part of the programme. Many urban roads are supported at the edges by cellars (generally but not always masonry or brick). A policy with these, too, has not been decided - neither has the responsibility of the owner and the highway authority. Records do not exist for these structures, considerations must be made in due course of the potential problem from these structures and it may be that a sample survey should be commissioned to determine the scale of the problem.

31. Load Testing Structures. The views of the Local Authority bridge engineers on the merits and advantages of load testing as a means of determining capacity are mixed. Half believe that regular load testing is a cost effective way of proving the carrying capacity of some bridges (28 out of 54). It is considered that further work must take place in

this field because load testing has a modest expense when compared with strengthening or reconstruction.

32. Parapet Strengthening. Parapets are a very important part of the structure or retaining wall. The Government proposal to bring parapets up to strength is clearly praiseworthy - particularly when the strengthening programme is taking place. (42 out of 54 Bridge Engineers agreed that it should be done at the same time as strengthening). However, a survey by CSS has shown that in some Local Authority areas up to 80% of bridge parapets on Local Authority roads are brickwork, ashlar or masonry. None of which at present meet any specific standard. The CSS have commissioned a research project to test the performance of brick, ashlar and coarse rubble parapets to determine their effectiveness as a parapet. If successful, this project will enable parapets to be designed in masonry and brick and will remove the necessity to consider replacing with metal parapets. This will have a potential saving in the order of £1b plus as well as very considerable environmental benefit. The CSS is pleased to have the support of the Department of Transport and the Welsh, Scottish and Northern Ireland Offices, as well as support from several Metropolitan District Councils and the British Waterways Board. The impact tests will be carried out in late 1993 and January 1994 and the author will hope to report on these during the presentation. It may interest people to know that up to 70-80% of all Local Authority bridge parapets are of brick or masonry construction as are almost all parapets to dry stone retaining walls.

CONCLUSIONS

33. i) The problem is likely to more severe (in financial terms) than at first anticipated;

ii) the programme is on target for Department of Transport bridges, but likely to be delayed on Local Authority structures;

iii) concern must be expressed at the late start of the Railway programme (excluding London Underground which is on programme);

iv) it is essential that Local Authorities attempt to complete the programme with full Government support or else the danger of having a heirarchy of first class, second class and third class roads (with respect to construction and use capacity) may occur.

v) the effect of Local Government reorganisation cannot be judged at present but inevitably it will make acceleration of the programme very difficult and is more likely to cause further delay.

vi) The programme is being very effectively managed by the Highway Authorities who need to set up sophisticated bridge management system in order to ensure a high level of condition of the country's bridges and to maintain that high level when the programme is complete. Bridges are a vital part of the country's infrastructure which must not be allowed to deteriorate.

34. I would like to thank all my colleagues in the County Surveyors Society and the Metropolitan District Councils for freely providing much useful information and comment on this very important subject.

REFERENCES (either implicit or referred to in the text)

35. Bridge Census and Sample Survey - DOT - 1987
New Concepts for the Management of Highway Structures - IHT - 1990
Bridge Durability - Symposium - Construction Marketing - 1990
15 year Rehabilitation Programme - DOT - TRM 2/90 App 2
Assessment and Strengthening of Highway Bridges and Structures Road 2/91
The Assessment of Highway structures - DOT- BD21/84, BA 16/84.

The Design of Highway Bridge Parapets - DoT - BD52/93
Bridge Maintenance Initiatives - IHT - April 1986
The Assessment of Highway Bridges and Structures - DoT
BD 21/93
BD 34/90
BD 44/90
BD 46/92
BD 50/92
and other associated subjects.

Avonmouth bridge: assessment and strengthening

J. GILL, Technical Director, Acer Special Structures, J. JAYASUNDARA, formerly Regional Bridge Engineer, Department of Transport, and C. COCKSEDGE, Technical Manager, Acer Special Structures

SYNOPSIS. The structural assessment and design of strengthening works for the Avonmouth Bridge is being undertaken in order to determine the works required to meet current loading standards including the requirement to carry 40 tonne lorries by 1999. The present structure carrying dual 3 lanes with hard shoulder and footway/cycle tracks needs strengthening to carry this loading. With the provision of some additional strengthening it is possible to rearrange the layout of lanes on the bridge deck to accommodate dual 4 lanes, hard shoulders and retain provision for pedestrians and cyclists. This coupled with other works on this short length of M5 motorway will reduce the traffic congestion currently occurring during peak periods between junctions 18 and 19. This additional capacity can be obtained within the present highway boundaries.

BACKGROUND

1. The M5 motorway is a strategic route linking the South West to the Midlands, Wales, London and the South East.

2. The Government's White Paper "Roads for Prosperity" and the "Trunk Roads, England into the 1990s" identified the length of the M5 between junctions 15 (Almonsbury) and 21 (Weston-Super-Mare) to be considered for widening.

3. The section of the motorway between junctions 18 and 19 passes west of Bristol over the Avonmouth Bridge and links Gordano on the southern bank of the River Avon to Avonmouth on the northern bank. The centre of Bristol is linked to the motorway via A4 Portway through the Avon Gorge.

4. The section between junctions 18 and 19 of the motorway carries the highest volume of traffic on the motorway network in the South West with gradients of 3.8% on both approaches to the bridge over the River Avon being one of the factors giving rise to congestion during peak traffic periods. The large proportion of local traffic using the motorway across Avonmouth Bridge is another factor and during the peak holiday periods volumes as high as 120,000 vehicles per day are commonly recorded. The combination of these factors and the

need to strengthen the bridge to meet current loading standards makes Avonmouth Bridge and this link in M5, a candidate for early attention.

5. Structural assessment of the bridge is substantially complete and design of the strengthening works is nearing completion.

DESCRIPTION OF AVONMOUTH BRIDGE

6. Avonmouth Bridge was opened to traffic in 1975. It was designed to British Standard BS153, but loading intensities have increased considerably since that time. Structures in this country are now required to carry the design loading specified in BD 37/88 which includes the 40 tonne lorries which will be admitted into the UK after January 1999.

7. The superstructure of the bridge comprises 20 spans of twin steel box girders with an overall length of 1400m. The spans vary between 30m and 174m. The boxes are about 6m wide and typically 3m deep but haunched to 7.5m at the main river piers. These main boxes are connected with steel cross girders and cantilevers at about 3m centres along the bridge to give a 40m wide deck. This deck is of composite concrete construction over the 1000m of approach spans, and steel orthotropic deck over the 400m of main span and side spans. A general arrangement of the crossing is given in Figure 1.

8. The superstructure is supported on pyramid shaped knuckle bearings fixed to piers that are designed to flex longitudinally. The piers are generally on spread footings except for the three just to the south of the river which are on piled foundations. The structure is fixed to each abutment and a rolling leaf type of movement joint is provided near the centre of the crossing.

9. The bridge superstructure weighs just over 40,000 tonnes of which about 12,000 tonnes is structural steelwork, 21,000 tonnes is reinforced concrete and 8,000 tonnes is surfacing.

10. The present carriageway configuration (illustrated in Figure 2) comprises of dual 3 lanes and hard shoulder, median and footway/cycle tracks on each cantilever. The original design assumed that two footway/cycle tracks would never be fully utilised and that the surplus space could be used for future widening if needed.

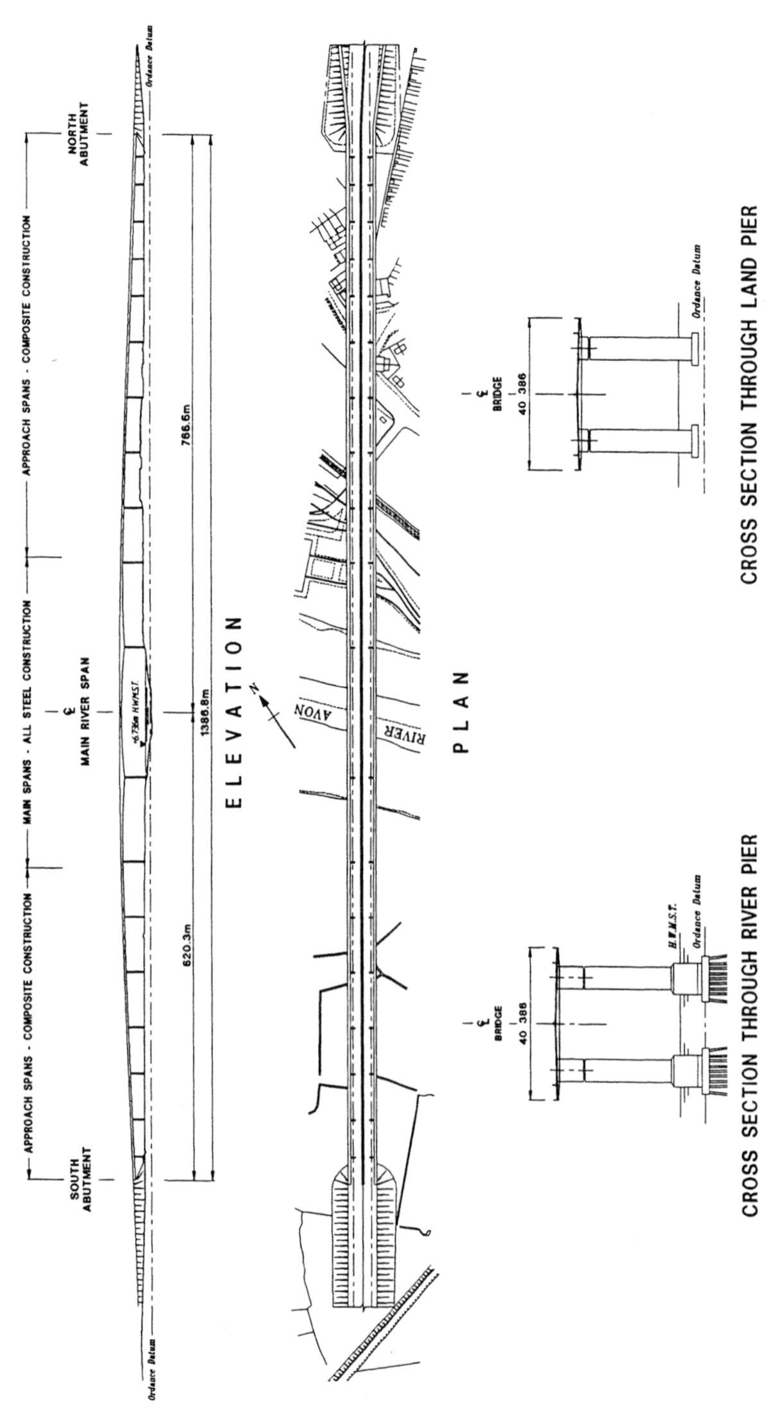

PART 2: CASE STUDIES

FIGURE 1

AVONMOUTH BRIDGE

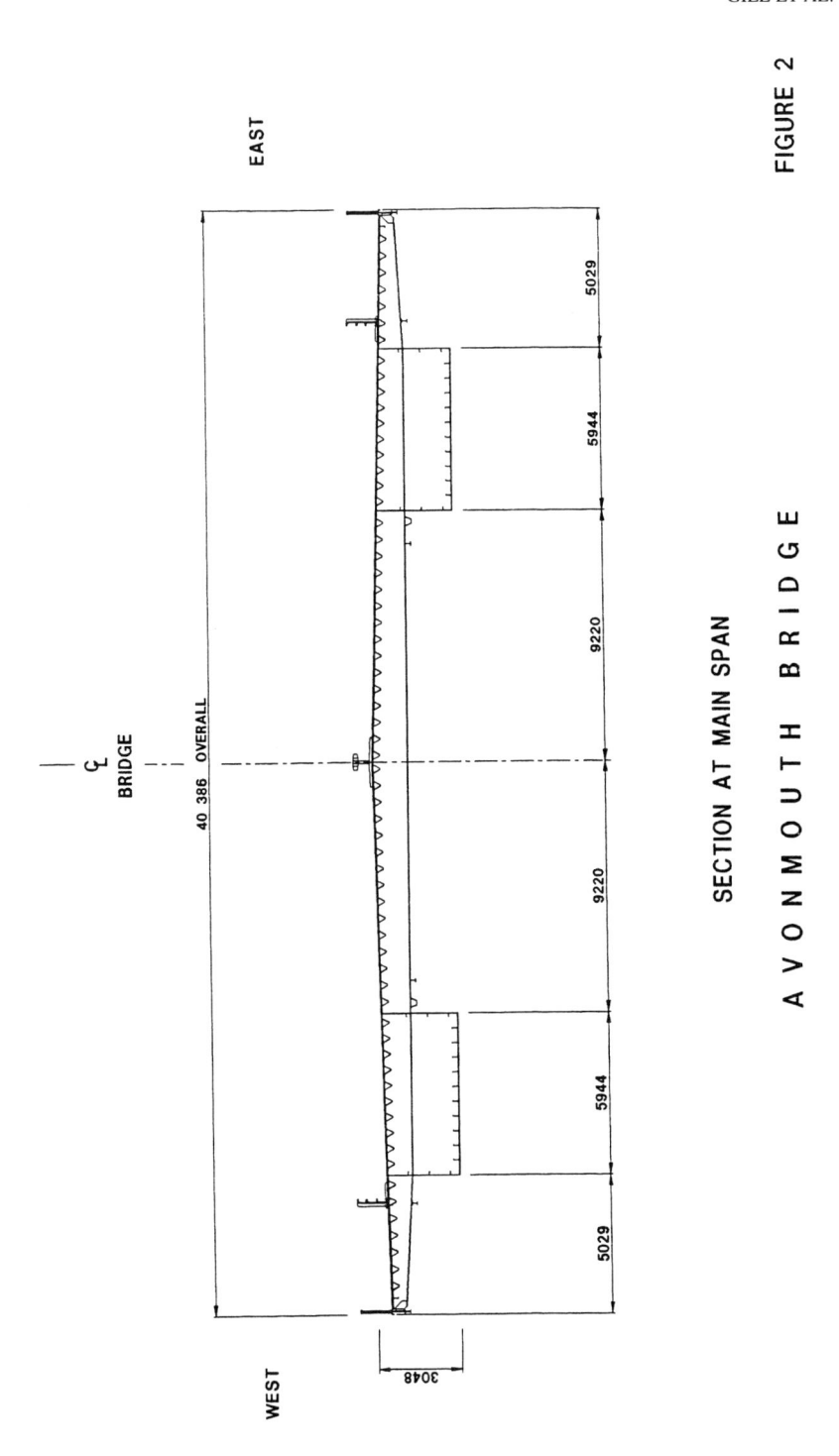

FIGURE 2

SECTION AT MAIN SPAN

AVONMOUTH BRIDGE

PART 2: CASE STUDIES

PROPOSED LAYOUT FOR STRENGTHENED STRUCTURE

11. It is proposed that the strengthened structure will have an asymmetric cross section maintaining the provision for footway/cycle track users on one side of the bridge. The lighting would be moved from the central median to the edges of the carriageway to make better use of the available space. The structure would accommodate dual 4 lanes, hard shoulders and footway/cycle track as illustrated in Figure 3.

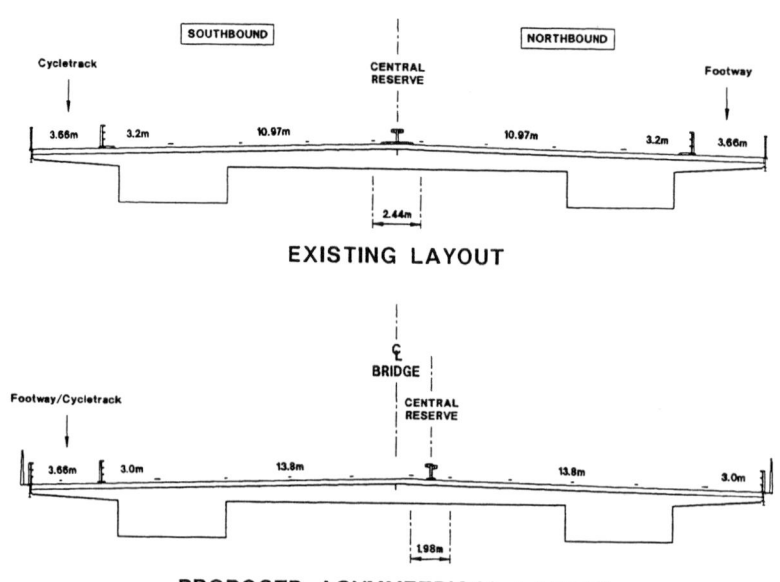

EXISTING LAYOUT

PROPOSED ASYMMETRICAL LAYOUT

FIGURE 3

BASIS FOR ASSESSMENT AND STRENGTHENING

12. Assessment and strengthening design was carried out for both the existing cross section (dual 3 lanes with hard shoulder and footway/cycle tracks) and for the proposed cross section (dual 4 lanes with hard shoulder and footway/cycle track).

13. Usual current DOT standards and guidance notes were used. However for some elements of the structure it was found that present design standards could be slightly conservative for assessing an existing structure of this size and complexity. We were able to define and develop, with assistance from the DOT BE Division, several departures from DOT standard which have yielded considerable saving in strengthening requirements.

DEPARTURES FROM DOT STANDARDS
Use of actual yield stress of box girder steel plate.

14. BS5400: Part 3 assumes yield stress for grade 50 steel of 355N/mm^2. This is an all encompassing value covering all steel made to BS 4360 (Weldable Structural Steels). In order to achieve the guaranteed minimum yield stress as well as other specified properties such as toughness, the actual yield stress will usually exceed the minimum values laid down in BS 4360. Under normal circumstances for a new bridge the designer does not have available the actual test results of the steel to be used and therefore uses the guaranteed minimum of 355N/mm^2. In the case of Avonmouth Bridge the majority of the mill test certificates for the steel plates actually used are available together with records of where the plates were used in the bridge.

15. The actual yield stresses of the steel plates used for the box girders on Avonmouth Bridge have been identified from the original mill certificates, and "maps" of the flange and web yield stresses produced for the whole bridge.

16. The nominal yield stress for assessment and strengthening of the bridge has been taken as:-
Nominal yield stress = Actual yield stress* - 10 N/mm^2
(*Identified from mill certificate.)

17. Each box unit is considered separately and is subdivided into top flange plate, web plate and bottom flange plate.

Control of HB vehicles and abnormal loads.

18. The proposed Departure from Standards is to impose the following restrictions on position on the bridge for heavy abnormal indivisible loads.

- a) Vehicle to travel down the centre of the box girder, guided by suitable road markings if necessary.

- b) No associated traffic will be permitted on the bridge in the same carriageway as the vehicle for a length of 300m in front and 300m behind the vehicle (approximately 2 spans).

- c) No special control over traffic in the carriageway not occupied by the vehicle.

19. The benefit of this restriction for Avonmouth Bridge is that heavy loads can be symmetrically distributed on the box girder. This has the effect of ensuring that the dominant loading for the structure is generally the equivalent uniformly distributed HA lane loading.

PART 2: CASE STUDIES

Reduction from BD 37/88 Footway Loading.

20. The existing footways/cycle tracks on Avonmouth Bridge are used regularly but at a very low intensity. Typically a total of about 180 crossings are made over a 12 hour day.

21. The nominal footway loading given in BD 37/88 is not realistic for global loading of footways on Avonmouth Bridge during normal usage (ie no crowd loading). However, it is possible that crowd loading could occur on the bridge due say, to a special event on the River Avon. In such a case the global loading could well reach the full intensity given in BD 37/88 but in this situation the use of the hard shoulder as a possible running lane will be restricted to emergency vehicles only.

Partial load factor to be applied to superimposed dead loading (surfacing).

22. On the composite approach spans of Avonmouth Bridge, the existing surfacing comprises of 19mm (0.75 in) mastic waterproofing overlain by 38mm (1.5 in) rolled asphalt surfacing, giving a total nominal thickness of 57mm surfacing.

23. On the orthotropic main spans, the existing surfacing comprises of 3mm (0.125 in) bitumen overlain by 38mm (1.5 in) mastic asphalt, giving a total nominal thickness of 41mm surfacing. Measurement of actual surfacing thickness has been undertaken using the ground penetrating impulse radar method.

24. As part of the carriageway widening and bridge strengthening works, the bridge will be completely rewaterproofed and resurfaced. The specified tolerance on thickness of surfacing for the whole structure for all future resurfacing work will be ±5mm for approach spans and main spans.

25. This design refinement uses gamma $_{fL}$ factors of 1.20 at ULS and 1.0 SLS derived from the theoretical study of the Specification and allowable tolerances in the thickness of steel, concrete and surfacing.

26. After a survey of present surfacing thicknesses it was evident that the factors given above are more than adequate.

Use of Interim Design and Workmanship Rules (IDWR/Merrison Rules) in assessing the orthotropic steel deck.

27. The main span of Avonmouth Bridge has an orthotropic steel deck which is longitudinally stiffened by Vee-troughs. When assessed in accordance with the requirements of BS5400: Pt. 3 the stiffeners were found to have a low strength capacity when acting as part of the compression flange. This was principally due to non-compliance with shape limitation requirements in the present standard, which require a reduced effective section to be taken and makes no allowance for the torsional stiffness of the closed stiffeners.

28. This refinement in design uses the Interim Design and Workmanship rules. This is permissible because the structure was fabricated to the IDWR tolerances rather than the less stringent requirements in BS5400: Pt. 6. Site surveys have been carried out to confirm that the actual tolerances achieved are within the requirements of IDWR. The use of IDWR is therefore beneficial when calculating the capacity of the trough stiffeners on Avonmouth Bridge.

Use of historic bridge temperature and thermal movement data to assess substructure restraint loads

29. This refinement of design uses historic data on temperature and thermal movement of the superstructure of Avonmouth Bridge to predict the maximum expansion and contraction which will occur during the design life. These maximum expansion and contraction values were then used to establish an average global coefficient of expansion for the bridge and hence the true restraint loads imposed on the substructure of the bridge.

Additional Considerations

30. In order to give added confidence in the calculated dead load stresses in critical areas of the bridge, the actual dead load stresses are being measured using the techniques employed for determining residual stress in structural members.

31. It should be noted that when strengthening an existing structure minimising the added strengthening steelwork has an important additional benefit. The structure is designed to carry both its self weight and the applied loads; by minimising the amount of extra steel added to strengthen the structure, the capacity for carrying traffic loading is increased.

PART 2: CASE STUDIES

RESULTS OF ASSESSMENT

32. The findings of the assessment indicate that strengthening is required in webs and flanges at piers and in flanges at mid-span sections. This is illustrated in Figure 4. Piers and foundations were however found to have adequate capacity both for the existing D3M and the revised D4M loading configurations, subject to confirmation from the results of a special ground investigation.

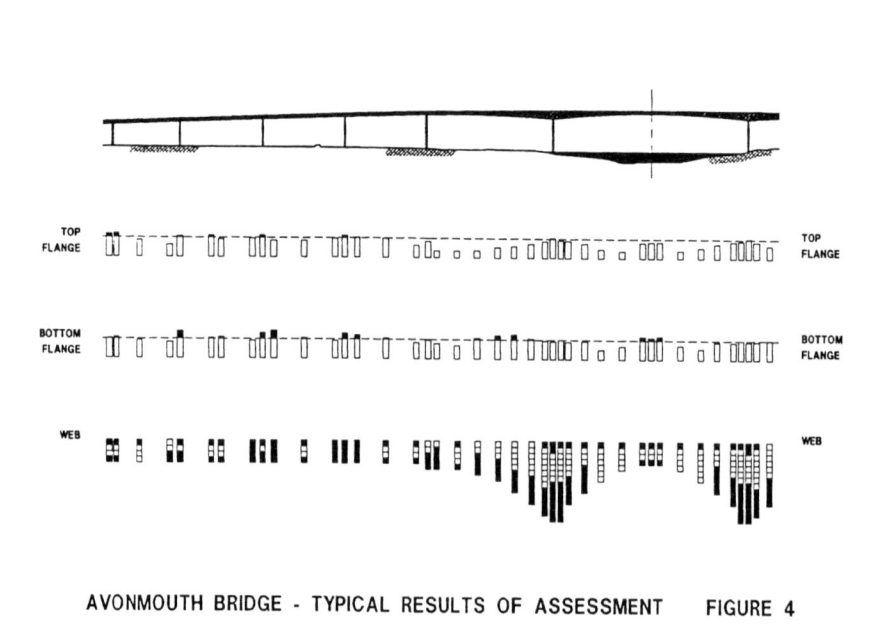

AVONMOUTH BRIDGE - TYPICAL RESULTS OF ASSESSMENT   FIGURE 4

DETAILS OF STRENGTHENING DESIGN CONCEPT

33. When designing the strengthening, the following basic principles were used:-

(i) Maintain 3 lanes of traffic in each direction at all times during the strengthening works.

(ii) Aim to keep all strengthening work within boxes.

(iii) Maximise the composite action of strengthening steelwork by making it carry original structural dead load where possible.

34. Temporary support is to be provided to relieve some original dead load before strengthening commences. Also strategic removal of superimposed dead load (surfacing etc) is to take place before strengthening is attached. Replacement of superimposed dead load will occur after strengthening becomes effective. This achieves better composite action.

Prestressing in approach spans. (Figure 5)

35. The prestressing system consists of a fan of bars anchored near the top of each pier diaphragm and near the bottom flange. This arrangement reduces the hogging moment at the piers and relieves some of the shear in the webs in this location, passing load directly into the diaphragm in an efficient manner. Temporary trestle supports will be used to support the boxes at the third points of two adjacent spans. The jacking load at each trestle position will be maintained at a constant level to relieve a proportion of dead load while the prestressing system is installed in one box. This procedure of trestle supports and prestressing is progressed along the length of the bridge.

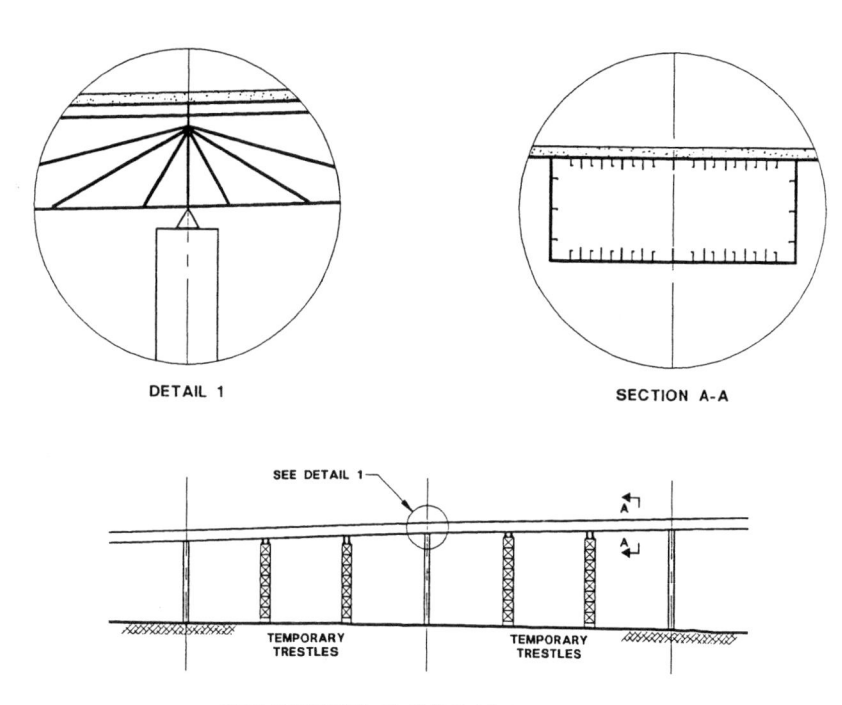

DETAIL 1

SECTION A-A

PART ELEVATION OF SOUTH APPROACH SPANS
(2No. SPANS TO BE JACKED AT A TIME)

AVONMOUTH BRIDGE - APPROACH SPAN STRENGTHENING   FIGURE 5

PART 2: CASE STUDIES

Strut and tie in main spans. (Figure 6)

36. In the main spans a different approach to strengthening has been adopted. At the main piers the boxes are significantly haunched and hence provide the space for an efficient strut and tie system to be installed. The system proposed comprises of circular hollow steel section struts laid on the bottom flange transverse stiffener reaching from the diaphragm to the third point of the span. A prestressing tie, is attached to the ends of the strut rising to pass over a saddle located near the top of the diaphragm. The stressing load applied to the ends of the strand will be monitored and adjusted to ensure that the required uplift effect is achieved at the third points of the span.

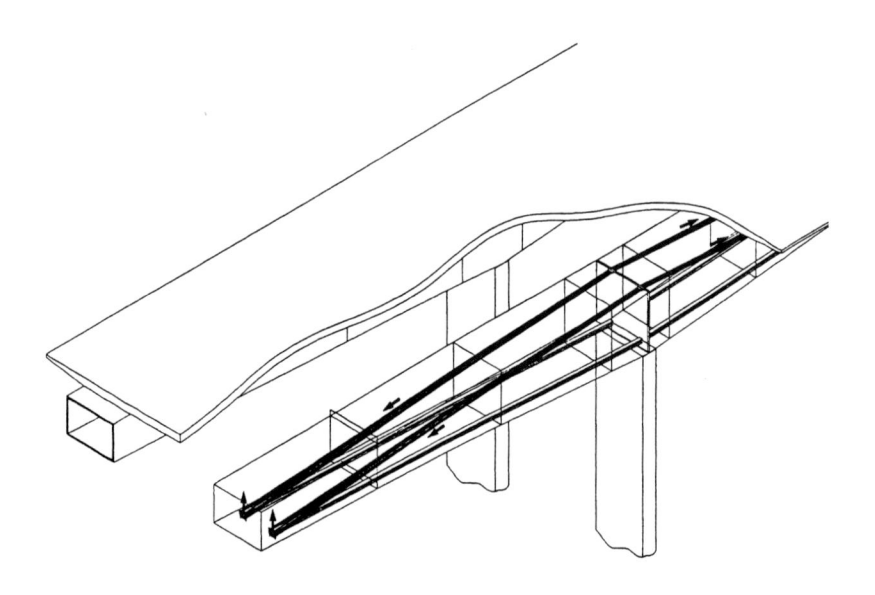

MAIN SPAN - PRESTRESSING   FIGURE 6

CONCLUSIONS

37. The requirement to strengthen Avonmouth Bridge to carry loading to modern standards as well as the provision of an extra traffic lane created a major engineering challenge.

38. If this had been treated as a new build requirement and full current design standards had been applied there would have been the need for the addition of some 6000 tonnes of steel to the existing 12,000 tonnes steel superstructure.

39. The ability to apply the various considerations outlined in this paper will result in significant savings in the amount of additional steel needed for the strengthening works. The final strengthened bridge will be structurally more efficient; the benefit in the reduction of strengthening in terms of cost, length of construction period and likely disruption to road users will prove to be invaluable.

ACKNOWLEDGEMENT

40. The authors would like to acknowledge the consent of the Director General of Highways of the Department of Transport to the publication of this paper.

Strengthening masonry arches

S. F. BROOMHEAD, Formerly Project Manager, and G. W. CLARK, Team Leader, British Rail Research

SYNOPSIS. British Rail Research (BRR) have conducted various studies into the behaviour of masonry arches. These studies have covered the assessment and repair of masonry structures. One such study involved a survey of arch repair and strengthening techniques employed by British Rail (BR) and this paper summarises the findings of this work.

INTRODUCTION
1. British Rail are responsible for around 33,000 arch spans carrying both rail over road and road over rail. Most of these structures are over 120 years old, the design life for modern structures, but are required to carry loads much greater than their designers could have envisaged. The ever-increasing weight of vehicles, especially road vehicles, means that many of these structures must be strengthened.

2. The masonry arch is a very durable structural form but it is difficult to analyse. Consequently, designers of masonry arches used 'rules of thumb' to proportion their structures. Recently, sophisticated arch analysis techniques have become available, for example the BRR 'MAFEA' computer package, and this has enabled the true capacity of arches to be determined. Frequently the arch has sufficient strength to carry modern loadings. All that is required is to ensure that the structure be in good condition. Thus the repair of an arch can amount to strengthening and in this paper no distinction is made between repair and strengthening works.

3. Arch defects can develop for many different reasons including weathering, ground movements, breakdown of waterproofing systems or the effects of traffic. Strengthening may be required due to these defects or because of an increase in loadings. In either case, an assessment of the structure will be required and the engineer must decide whether to repair the structure. There are two basic reasons that will lead to repairs being carried out. The first is if a fault is assessed as directly affecting the arch in terms of limiting the live load. The second reason is that although the fault may not currently restrict the live load capacity of the bridge, failure to repair the fault may lead to further more serious damage developing. The correct identification of the reasons for repairing the

structure is important because it governs the criteria by which the success of the repair should be measured.

4. In this paper, the subject of arch assessment will be briefly discussed as it is the process by which the capacity of an arch is calculated and therefore the process that limits traffic over the structure. The various repair techniques used by BR will be dealt with in turn. General advice on the repair and maintenance of masonry bridges is contained in reference 1.

ARCH ASSESSMENT

5. Arch assessment consists of the calculation of the load capacity of the arch barrel and also the formation of a subjective opinion of the stability of the arch's spandrel walls.

6. The most commonly-used method of arch assessment is the modified MEXE method described in the Department of Transport advice note BA16/93 (ref. 2). With Modified MEXE Method assessments the capacity of a parabolic arch in perfect condition is initially calculated. This is modified by the application of factors to account for differences in geometry, materials and faults between the real arch and the ideal structure on which the calculation was based. The method is quick and simple to apply but suffers from several disadvantages:-

(a) The safety margin (the relationship between the permissible load and actual arch collapse load) is unclear.
(b) The method is inflexible, preventing modification by the assessment engineer to account for different situations, for example arches with haunching.
(c) The use of crude factors to take account of arch geometry, materials and faults is obviously not entirely satisfactory.

7. Various computer programs have been developed to improve the assessment of arch barrels. These programs are not as quick to use as the MEXE method as they require more information about the structure than MEXE does. Nevertheless, an initial assessment can be performed quite quickly using conservative assumptions of material properties. A range of values of the unknown properties can be used to study the relative importance of each parameter. This allows the important variables to be identified and site investigations can be targeted on them.

8. Arch assessment is not only concerned with the assessment arch barrel condition but also the assessment of spandrel and parapet wall condition. At present this assessment is entirely subjective, but research is underway into assessment programs able to deal with spandrels.

ARCH REPAIR SELECTION

9. Various factors can influence the choice of repair technique. The following factors are important with regard to BR

PART 2: CASE STUDIES

bridges (in no particular order of priority).

- (a) Type of fault that is to be repaired.
- (b) Ease of access.
- (c) Limitations on the appearance of the repaired bridge.
- (d) Available clearances.
- (e) Length of possession time available.
- (f) Costs of the various repair options.
- (g) Expertise required to carry out repair.
- (h) Length of time that repair is required to last.
- (i) Purpose of repair, (strengthening or prevention of further damage)
- (j) Obstruction of future arch barrel inspections.

10. Type of fault that is to be repaired. This is obviously a major influence on the selection of repair method. In many cases the repair scheme will aim to repair several faults, for example a concrete saddle combined with tie bars and a waterproofing membrane may aim to repair a cracked arch barrel, detached spandrels and inadequate waterproofing.

11. Ease of access. Bridges, by their very nature, span obstacles and this can make the provision of access difficult. For underbridges access to the intrados of high viaduct spans or river spans is difficult and for overbridges scaffolding cannot be used where it might infringe on the structure gauge.

12. Limitations on the appearance of the repaired bridge. This limitation can be a formal requirement, as with a listed structure, where repairs that degrade the appearance of the bridge are not tolerated, or it may be left to engineering good practice to design an aesthetically pleasing repair.

13. Available clearances. Overbridge repairs, carried out on the intrados of the arch, must not interfere with the railway structure gauge. For underbridge repairs over a road, the repair must not restrict the headroom that BR is obliged to provide.

14. Length of possession time available. The time available in which to repair a structure is not unlimited (either over roads or railways) and therefore it may not be possible to carry out repair methods that would require a long time to do.

15. Costs of the various options. The cost of various repair schemes clearly influences the choice of repair method, but it may not be easy to identify all costs associated with some repair methods at the outset.

16. Expertise required to carry out repair. If the skills required to carry out a particular type of repair are not widely available then the engineer may be more reserved towards this technique.

17. Length of time that repair is required to last. Repairs fall into two categories, those that are designed to be relatively permanent and those that are only designed to be a temporary measure because the structure is likely to suffer further damage or is to be reconstructed shortly. One example is the use of steel ribs to repair arches suffering from mining

subsidence.

18. <u>Purpose of repair</u>. The purpose of the repair also governs whether a repair is appropriate. For example, a tie bar and pattress plate repair on the spandrel walls only constrains further movement of the spandrel walls, it does not strengthen the arch barrel.

19. <u>Obstruction of future arch barrel inspections</u>. Repairs carried out below the intrados of the arch can obstruct the future inspection of the arch barrel. If the repair is substantial, such as a thick cast insitu concrete lining then inspection to determine the condition of the arch barrel will not be required. If however the repair is of a lighter nature, such as a thin sprayed concrete layer, then the condition of the arch will still be important.

ARCH REPAIR TECHNIQUES

20. Some of the many repair techniques employed by BR will now be examined. A brief description of each repair technique is given along with installation details. The purpose of the repair is considered and the effect of the repair on subsequent assessments of the arch is discussed as well as any recommendations concerning the particular repair method.

21. <u>Spandrel tie bars</u>. This is one of the most common repair techniques and has two variants: Pattress plates can be fixed to each end of a tie bar which passes completely through the structure or a single pattress plate can be attached to a bar which is anchored in the structure but does not pass completely through it.

22. The purpose of the repair is to stabilize the arch spandrel walls which often push outwards due to horizontal pressure exerted by the fill. These forces are attributed to the spread of the fill, increased ballast depth (rail bridges) or increased live load.

23. Tie bars are installed by either drilling through the arch and fill or by being laid on the backing after excavation from above. The tie bars are tightened only nominally as their function is simply to restrain further movement. The positioning and number of tie bars is governed by the extent of the particular fault but very strong consideration should be given to the appearance of the bridge and every effort should be made to maintain the symmetry of the structure. In cases where the spandrel brickwork is weak, spreader beams may be used to distribute the tie bar load over a wider area. If the structure is listed, the use of tie bars and pattress plates can be difficult because of aesthetic requirements. A possible solution is to rebate the pattress plates into the spandrel walls.

24. As the purpose of the tie bars is to prevent the spandrels from moving outwards and the assessment method only considers the arch barrel in strength calculations, it is difficult to allow for the effect of tie bars in the arch capacity calculation.

25. <u>Spandrel strengthening</u>. This technique consists of casting a reinforced concrete beam behind the spandrel to give a

composite spandrel of greater mass and stability than the original. A trench is excavated behind the spandrel wall down to the back of the arch barrel. Ties are then attached to the spandrel wall and reinforcement fixed. Concrete is poured in lifts of a maximum one metre depth to prevent the hydrostatic head of the wet concrete pushing the spandrel wall off.

26. The purpose of this repair is to strengthen the spandrel walls, therefore if the engineer is confident that the repair has been successful, any speed and load restriction previously due to the poor spandrel condition could be lifted. This type of repair may be particularly suitable when strengthening the parapets of a bridge.

27. Saddling. A saddle consists of concrete placed on the extrados of the arch barrel to increase the strength of the arch. The purpose of a saddle is to repair an arch that is suffering from arch barrel faults often combined with waterproofing problems and spandrel wall faults. The saddle is distinguished from the relieving slab in that it is applied throughout to masonry, either to the extrados of the barrel or sound haunching whereas a relieving slab is applied directly onto the fill above the arch - ie saddles are curved, relieving slabs flat.

28. Arch saddles can be designed in several different ways. The new saddle may be designed as a reinforced concrete arch capable of supporting the full dead load and live load. The only requirement of the existing arch barrel is to act as a former for the new concrete arch. This new concrete arch does not act in the same manner as the old masonry arch, the reinforced concrete is capable of supporting tensile stresses so the concrete arch saddle is effectively a curved beam and not a gravity structure reliant on the abutments to carry thrust.

29. An alternative design approach is for the saddle to contain nominal reinforcement only (to prevent excessive cracking of the concrete during the curing process). In this case the saddle acts only in compression and the new composite arch can be regarded as an arch with a total thickness equal to that of the barrel plus saddle. However, since the saddle acts in the same way as the arch, it imposes horizontal forces on the abutments. Therefore if this design approach is adopted, provision for the support of these horizontal forces should be made by keying the saddle into the haunching and abutments.

30. The majority of saddle designs tend to behave in a way which lies somewhere between these two extreme design philosophies. The drawback to this is that the precise nature of the repair is not apparent and this could lead to assessment problems in the future. The early passage of vehicles across the structure should be avoided as this can lead to cracking of the repair before it has had sufficient time to develop its working strength.

31. The exact proportion of the live load that the repair is designed to carry varies as discussed above. It can be used to provide a new reinforced concrete arch of the required strength where the function of the old arch is reduced to that of

permanent formwork. Alternatively, a saddle can be provided to increase the thickness of the arch barrel so that the strength of the composite (concrete/masonry) arch is increased. The main advantage of the saddle is that it can easily be combined with re-waterproofing and spandrel wall repairs.

32. The manner in which the structure should be assessed after the repair depends on the design philosophy that has been adopted. If the sole function of the old arch becomes cladding, then so long as the masonry can support its own dead load and is in no danger of falling out, the arch will be deemed to be satisfactory. However if a much smaller saddle has been provided to increase the arch strength by the provision of a composite barrel consisting of both the original masonry and the new concrete, and no reliance has been placed on the tensile capability of the concrete, then the assessment of the arch should reflect this. In its simplest form the arch would be assessed using existing masonry arch assessment codes such as the Modified MEXE Methods based on the increased barrel thickness.

33. There have been instances of the masonry arch barrel deteriorating rapidly after an arch has been saddled. The precise mechanism involved in these failures is not fully understood, but it may be related to the greatly different stiffnesses of masonry and concrete and the ability of the reinforced concrete saddle to carry bending stresses.

34. Relieving slabs. A relieving slab is a reinforced concrete slab placed over the arch after the fill has been partially excavated. Relieving slabs are often built with side walls so that a shallow trough is formed thus reducing the spandrel wall loadings. This method is applied normally in circumstances where it is difficult to remove totally the fill. This may occur on deep semi-circular viaduct spans for example. The installation will be similar to that of the saddle and will normally be combined with waterproofing and spandrel wall repairs.

35. The aim of a relieving slab is to reduce the live loading on the arch. The relieving slab acts as a beam, spanning from abutment to abutment. The reaction at the abutment is therefore vertical. This gives a loading that is different from a conventional masonry arch which generates horizontal and vertical reactions on abutments.

36. A problem with this type of repair is that the repaired structure can no longer be regarded as either an arch or a simple beam. This may make future assessments complex. As the stiffness of a relieving slab is increased, it will carry a greater amount of the live load. At the limit, when it is infinitely stiff, all the loads will be carried by the relieving slab and transferred vertically to the abutments. In this case, deflection measurements of the arch barrel would show no deflection under traffic. Conversely if the slab is made too flexible then it will not affect the arch, and deflections will remain unchanged after the repair.

37. A possible means of assessing the repair is to measure deflections under a given service load before, and following

the repair. If the deflections were the same, the effect of the relieving slab is zero and the arch must continue to be assessed as having to support the full live load. If the measured deflections were zero then the relieving slab would have been totally effective in relieving the arch and the arch need only be assessed as having to be self supporting. In between these two extremes the live load would be shared pro rata.

38. This type of repair can, when construction depth is limited, lead to a high loading on the arch crown due to the relatively stiff nature of the relieving slab compared to the flexibility of the fill. Some evidence, of increased crown damage has been found to support this concern.

39. Waterproofing. Waterproofing is one of the most important arch repairs, as without a satisfactory waterproofing layer and efficient drainage an arch can deteriorate very rapidly.

40. New waterproofing systems can be divided into three categories, the first is the High Level system and is installed between the fill and the ballast. The second is a Low Level system and is installed between the fill and the masonry barrel. The third is when the fill itself is waterproofed by the use of chemical grouts. The waterproofing of the arch intrados is not recommended as this would ensure that the masonry is permanently saturated and susceptible to spalling in cold weather.

41. The High Level system is the most preferable, as this will ensure that the fill is maintained in a dry state and this in turn will reduce spandrel wall damage due to hydrostatic pressures or the swelling of the fill. As waterproofing is normally undertaken when 'above extrados' repairs such as saddles and relieving slabs are being installed this type of waterproofing system is the automatic choice.

42. The Low Level system is rarely used as it offers no advantage over the High Level system and has the possible disadvantage that greater excavation may be required.

43. The third system of waterproofing, using chemical grouts to waterproof the fill, is rarely used but has the advantage that a waterproofing system can be installed without disruption to the track and ballast.

44. This repair would not affect the assessment of arch strength. Its purpose is the prevention of further damage.

45. Brickwork/masonry repairs. This classification of repair includes repointing damaged mortar, patch brickwork repairs (where small patches of damaged bricks are replaced) and re-ringing. However the replacement of a complete ring of brickwork is becoming increasingly rare due to the high costs when compared with sprayed concrete.

46. The purpose of a brickwork repair is to remove damaged or loose brickwork and mortar and replace it with sound brickwork. This will not only restore the strength of the arch but will also prevent an acceleration of damage. It is important that the materials used to repair an arch should be as close in terms of mechanical properties, appearance and colour match to the original as possible.

47. As the purpose of this repair is to remove damaged bricks

and replace lost mortar, it will have a slight effect on the capacity of an arch, tending to restore it to its original condition.

48 <u>Grouting</u>. Grouting consists of injecting a cementitious or resin fluid into the cracks and voids in the arch barrel. It is used to repair voids within the barrel and to repair barrel cracks such as ring separation. However if the cracking is extensive it is likely that stitching would also be necessary. There are many different forms of grouting, but by far the most common technique employed by BR is pressure grouting. In this method a matrix of holes is drilled into the structure and flushed out by water, prior to injection pipes being inserted. The grout is then injected into the structure at the lowest point and gradually progresses upwards. At each point, injection continues until the point will not accept any more grout or the grout emerges from a neighbouring hole.

49. A limitation of pressure grouting is the maximum pressure that can be applied to inject the grout. If the pressure is too high the possibility of blasting off a weak masonry face exists. Both cementitious and resin grouts are used for pressure grouting. One of the greatest difficulties associated with grouting is ensuring that the grout reaches or stays in the intended area.

50. Resin grouts can be manufactured to give a wide range of material properties, however they are more expensive than cementitious grouts and as a result are not used as extensively.

51. The premature passage of vehicles across the structure after grouting should be avoided as this can lead to cracking of the repair before it has had sufficient time to develop its full strength.

52. In the Modified MEXE Method factors are used to account for the reduction in arch capacity resulting from arch barrel cracks. If the engineer has confidence that the integrity of the arch has been improved by grouting it does not seem unreasonable that the appropriate factor could be increased to reflect this. As the structural function of the barrel is not altered by the repair, the use of before and after measurements is ideal for assessing the effectiveness of the repair. It is suggested that simple deflection tests are performed on the structure before and after the repair. A decrease in arch deflections for a given load shows that the repair has been significant in terms of the capacity of the arch.

53. <u>Stitching</u>. Stitching of the arch barrel consists of grouting dowels into holes bored into the arch barrel to restore its integrity. The technique is used where the arch barrel damage is beyond the scope of a simple grouting repair. Stitching is of particular use for the repair of large cracks and ring separation as the steel dowel will allow shear transfer across the cracks.

54. A pattern of approximately 20 mm diameter holes are initially drilled in the brickwork region requiring repair. Stainless steel or galvanised steel ribbed dowels of approximately 12-16 mm diameter are then inserted into the

PART 2: CASE STUDIES

holes, the dowels can be slightly curved to enable them to grip the sides of the hole. The dowels are then grouted into position, normally using a cementitious grout. The early passage of vehicles over the structure can cause damage to the repair as the grout will not have had sufficient time to develop adequate strength.

55. <u>Sprayed Concrete</u>. Sprayed concrete has been used by British Rail for arch repairs for many years, the process involves the placement of concrete by a spray nozzle without the requirement for extensive formwork on the arch intrados. The purpose of the repair is to strengthen the arch by the formation of a composite arch consisting of masonry and concrete layers.

56. Sprayed concrete can be placed by three different processes. The first is the dry process in which the dry cement and aggregate mix are transported to the nozzle by a high velocity air stream, where the operator controls the introduction of water. The second process is the wet process in which the cement, aggregate and water are mixed and then pumped to the nozzle where air is finally introduced. The third process is called the composite process and it was developed by the Hungarian Dorog Mining Organisation. The cement, aggregate and water are mixed to produce a very dry concrete which is transported to the nozzle where air is introduced and extra water.

57. Both the wet and dry processes suffer from problems. In the dry process there can be a wide variation in concrete quality as it depends greatly on the operator's skill as he both controls the mixing process and the placement of the concrete. The wet process is difficult to apply overhead (which is essential for arch repairs) and shrinkage can be a problem.

58. The composite process used by BR offers advantages in concrete quality and has been used to repair arches in most Areas of BR. The repair can be used to strengthen the arch by the addition of a concrete layer to the intrados (the minimum recommended thickness is 150 mm). If clearances are a problem and the bottom brick ring is in poor condition then it can be removed and replaced by a sprayed concrete layer. The loose mortar and brickwork should initially be removed. Steel bars are then resin grouted in to the existing arch prior to the concrete being placed to ensure the concrete repair and existing brickwork act homogeneously. Anti-cracking reinforcement is also required to prevent cracking during curing.

59. As the concrete carries only compressive stress it should act compositely with the existing arch structure. The assessment of the strength can therefore be based on the combined thickness of the concrete and masonry.

60. Although the repair can be applied to an underbridge without a possession, the arch deflections under traffic could crack the new repair before the concrete has developed sufficient strength and prevent bonding between existing arch and concrete repair. The repair should therefore only be carried out during a possession. Sprayed concrete should not be used for the repair of brickwork patches. It should only be used to

provide a complete lining of concrete on the intrados.

61. <u>Cast in-situ concrete soffit lining</u>. This repair consists of casting a concrete lining beneath the arch intrados. The concrete lining normally contains substantial reinforcement. The lining can be either from the springing or foundation level. The purpose of this repair is to strengthen an arch suffering from major defects of the arch barrel, such as diagonal cracks or distortion.

62. The concrete is placed over custom-made formwork, often from above via a hole through the existing arch barrel. After the concrete has set, any shrinkage cracking that has occurred between the repair and barrel is grouted up. This repair is normally designed to carry the entire live load.

63. The repair is designed to carry the entire live load then there should be no restrictions on the arch after repair.

64. <u>Steel plates or corrugated sheet lining</u>. This repair differs from the previous one in that the formwork is permanent. Preformed flat or corrugated steel sheets are used as formwork. Grout is then injected into the gap between the lining and the barrel. This repair is applied to arches suffering from brickwork deterioration. The repair improves the integrity of the arch by confining the arch barrel and preventing loose masonry from becoming detached. The grout introduced will also repair cracks in the brickwork, so strengthening the structure. The corrugated steel lining will not be structurally significant on its own but the effect of 12 mm steel plate should enhance the arch strength. This repair has the advantage that the reduction to bridge headroom is very small.

65. The repaired arch should be assessed based on the original arch thickness, as the extra thickness of the grout will be small. The factors that account for arch condition could be increased if the engineer has evidence that the arch has been strengthened. A simple deflection test can be used to indicate the effectiveness of this repair.

66. The corrugated steel plate repair is not very pleasing to the eye so should not be used for permanent repairs or on structures where the appearance is important.

67. <u>Steel Arch Ribs</u>. This repair consists of the provision of steel ribs to support the intrados of the arch. The method is most appropriate for the repair of an arch suffering damage resulting from ground movements. The repair permits the arch intrados to be inspected and if further movements occur the packing between ribs and arch can be adjusted to accommodate these.

68. For this type of repair to be used, the arch condition will be extremely poor making accurate assessment of the arch very difficult. Normally the steel ribs will be designed to be sufficient to carry the live load. The ribs naturally reduce clearances and can only be applied where sufficient clearances exists.

PART 2: CASE STUDIES

REFERENCES
1. SOWDEN A.M. The Maintenance of Brick and Stone Masonry Structures. E & F.N.Spon, London, 1990.
2. DEPARTMENT OF TRANSPORT The Assessment of Highway Bridges and Structures. Advice Note BA 16/93, London, 1993.

Bridge strengthening with minimum traffic disruption

B. PRITCHARD, Consultant to W. S. Atkins Consultants Ltd and Colebrand Ltd

SYNOPSIS. The cost of traffic disruption during bridge strengthening can readily exceed the strengthening costs themselves. The paper describes some bridge strengthening procedures which minimise, or even dispense with the costs and irritations of traffic disruptions.

INTRODUCTION

1. An ever-expanding world demand for highways means a global stock of bridges not only increasing in numbers but also in age and damage from road salt, air pollution and ASR. At the same time, traffic and the bridge loading demands of design codes are also increasing, leading to associated expansions of bridge strengthening programmes. In the UK one programme alone, to cater for the European Community requirement of heavier 40t lorries by 1999, will cost over £1 billion.

2. Bridge strengthening inevitably requires changes to the deck and/or substructure piers and abutments, often accompanied by repairs and upgrading. This means disruption to traffic carried and, in many cases, traffic crossed. Notional costs of such traffic disruptions can be high. Costs for providing 2+2 lane contraflows on a disrupted dual 3 lane motorway can exceed £200,000 a week. It is not surprising therefore that traffic disruption costs during bridge strengthening can well exceed the strengthening costs themselves. For this reason a considerable amount of ingenuity has been expended by bridge designers in devising bridge strengthening procedures which minimise, or even dispense with the costs and irritations of traffic disruptions.

3. A listing of the special requirements sought of such strengthening procedures is followed by descriptions of several minimum traffic disruption strengthening methods used for improving deck bending, deck fatigue life and substructure capacities.

BRIDGE STRENGTHENING REQUIREMENTS FOR MINIMUM TRAFFIC DISRUPTION

4. To minimise or even avoid disruption to the traffic carried or crossed during the bridge strengthening operations, the following characteristics are desirable:

PART 2: CASE STUDIES

- strengthening work to the deck should take place under the deck and away from the trafficked upper surface. There should be sufficient headroom for these works, although temporarily reduced headrooms are often permitted
- strengthening should be confined to adding to the existing structure, with only minor cutting or removal to acceptable safety factors
- strengthening materials should generally exclude wet construction such as in-situ concrete, guniting, grouting or gluing to avoid possible separation effects due to traffic vibration during setting
- strengthening attachments to existing steel bridge decks should be by bolting rather than welding and the attendant dangers of overheating steel under service loading
- strengthening procedures involving the use of relieving loads should not overstress the existing structure during application
- if added prestressing is used for deck strengthening, the deck will undergo new rotations and longitudinal movements. If is therefore essential to ensure that deck bearings and expansion joints can accommodate these effects. Excessive bearing friction can significantly reduce the prestressing input (Reference 1).

5. The strengthening methods described are technically feasible without resorting to any disruption of the traffic carried. However, some of the operations do require drilling into or applying prestress to decks subject to traffic live load. Due care will be required, preferably backed up by quality assurance procedures.

6. With acknowledgement to the various Murphy's Laws, some engineers may consider it prudent to close the deck to traffic whilst drilling or prestressing is taking place. Fortunately, these are short term procedures which can usually be accommodated during overnight closures.

DECK BENDING RELIEF STRENGTHENING BY PRESTRESSING

7. Conventional prestressing of a bridge deck imposes a permanent direct compression together with a bending moment which counters the dead and live load moments applied to the deck. The bending moment reduction effect of added prestressing can also be used to advantage in relieving bending in existing overloaded decks. This bending relief can be sufficient to reduce the deck bending under full dead and live loading to permissible limits. Alternatively, a bridge deck can be upgraded to carry increased superimposed dead and/or live loading equivalent to the dead load bending relief imposed. In most cases the magnitude of the bending relief is limited to the dead load only, as anything greater could cause undesirable bending tensions in the deck in the unloaded condition. It should be added that strengthening by prestressing can also provide extra shear strength by tendon inclination if required.

8. In general, the direct compression effect of the added prestressing is not as helpful as the bending relief effect. In reinforced concrete decks, allowable compressive stresses are usually lower than with prestressed concrete. In the steel girder elements of composite decks the extra compression can lead to local plate stability problems. With the bending relief required for the strengthening equating to the product of the added prestressing load and its eccentricity from the deck section centroid, there is every advantage in locating the prestressing tendons at the beam extremities or even beyond. By thus maximising the eccentricity, the required prestressing load is reduced to a minimum, with accompanying reductions in unwanted compressive stresses and, indeed, strengthening costs.

Strengthening a Reinforced Concrete Bridge Deck by Prestressing

9. A major trunk road crossing of the River Tawe was built in Carmarthen in 1938. The river was spanned by three reinforced concrete arches and the road and railtracks running along the river banks were crossed by a series of reinforced concrete beam and slab approach spans.

10. After 40 years use it became necessary to strengthen the whole crossing to cater for the increased traffic and newer bridge code requirements. The nature of the crossing required minimum, if any, traffic disruption. Several innovative techniques were developed (Reference 2), including the building of new reinforced concrete arches inside the existing arches, which were used as falsework, and the use of steel portals to strengthen some of the original beam and slab approach decks.

11. Self weight relief strengthening was applied to a three span section of the approach viaducts. The continuous 8 m beam and slab spans crossed the riverbank road and railway and the strengthening method adopted required no disruption to either the traffic carried or crossed. Strengthening was accordingly carried out under the deck by external prestressing to reduce self weight bending sufficient to allow the imposition of extra live loading without increased overall deck bending.

12. The prestressing anchor and deflector plates were fixed to the opposite sides of each beam by clamping on to epoxy mortar bedding using HSFG bolts drilled through the beams, Figure 1. The bolts were also grouted with an epoxy mortar to ensure that the prestressing force was transmitted from the anchorages to the beam by the twin actions of end bearing of the bolts and friction and adhesion generated by the combined HSFG clamping action and epoxy bedding.

13. The prestressing cables were 15 mm diameter 'Dyform' strands encased in protective plastic sheathing, Figure 2. No further protection was provided for these external tendons though provision was made for future topping up of the tendon loads if found necessary.

PART 2: CASE STUDIES

**Figure 1
Anchor Plate Fixing**

**Figure 2
Prestressing Strengthening Completed**

Strengthening of Steel/Concrete Composite Bridge Decks by Prestressing

14. Rakewood Viaduct, located adjacent to the Lancashire/Yorkshire boundary carries the dual 3 lane M62 motorway across a 36 m deep valley, Figure 3. The 34.4 m wide viaduct was constructed between 1967 and 1969 and is a six span continuous structure with two end spans of 36.6 m and four main spans of 45.7 m - an overall length of 256 m. Construction is of braced continuous steel plate girders 3.05 m deep, composite with an in-situ reinforced concrete deck slab.

Figure 3
Original M62 Rakewood Viaduct

15. The steep M62 gradients coupled with the high percentage of commercial vehicles using the motorway resulted in severe congestion of the two nearside lanes by heavy vehicles, with cars effectively restricted to the offside lane. Additionally, the site is very exposed, can suffer high winds and, in winter, severe snow and icing. To alleviate some of these problems it was decided that a climbing lane and a new hardshoulder should be provided for this section of the eastbound carriageway. Rakewood Viaduct lies within this climbing section and, to avoid extensive viaduct widening, it was further decided that the present hardshoulder across the viaduct should be used as the fourth eastbound running lane, the new hardshoulder being discontinued for this short length.

16. Inspections and assessments undertaken in 1986 and 1987 indicated that the viaduct had stood up well to the rigorous environment and required only minor salt corrosion repair treatments. However, the structure, originally designed to BS 153, was assessed to the current BS 5400 bridge code. The increased live loading associated with BS 5400 together with the road layout change, from 3 lanes plus hardshoulder to 4 lanes, meant

PART 2: CASE STUDIES

heavier design loading on the structure. The main shortfall was identified as an approximate 40% overloading in the steel girder compression flanges over the piers.

17. The overstressing of the bottom flanges over the piers by the proposed upgrading was complicated by the fact that the BS 5400 assessment also showed that the flanges were almost at allowable stress level under deck self weight alone. In the circumstances, it was decided to adopt the permanent unloading technique by prestressing for the required deck strengthening (Reference 3).

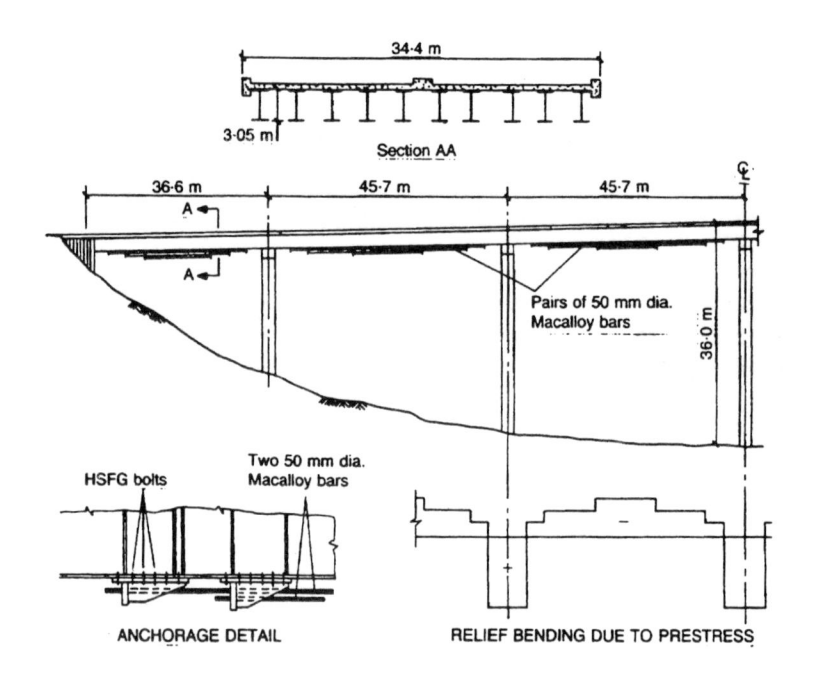

Figure 4
Rakewood Viaduct Strengthening Procedure

18. Figure 4 indicates the procedure, which first requires the attachment of fabricated steel anchors to the underside of each steel beam bottom flange by HSFG bolting. Three pairs of 50 mm or 36 mm diameter Macalloy prestressing bars of overlapping lengths are then attached under each flange between piers. Upon stressing, hogging bending is set up in the mid-span regions of the beam. However, it is the parasitic sagging moment over the piers, caused by deck continuity, which performs the required unloading to acceptable stress limits in the bottom girder flanges over the piers.

190

Figure 5
Prestressing Tendon Anchor

19. The dispersion of the high anchorage loads, Figure 5, into the girder flanges and webs, and the associated design of the extra local web stiffeners had been examined using three dimensional finite element techniques. Special consideration was also given to the provision of node supports to prevent wind vibration of the stressing bars and, of course, anti-corrosion protection. The prestressing of the girders proceeded with minimum disruption to the M62 traffic. The downstanding overlapped Macalloy bars and anchorages, Figure 6, caused little visual intrusion to the deck.

Figure 6
Strengthened M62 Rakewood Viaduct

20. In common with the earlier American work, it was found that the strengthening prestressing loads could be significantly reduced by excessive substructure restraint. It is therefore essential to ensure that the deck support bearings can articulate freely in response to the newly imposed movements due to the added deck prestressing.

21. It is believed that the technique could be improved further by substituting high strength fibre composite prestressing tendons for the steel Macalloy bars. Not only would the tendons require no corrosion protection, but their fatigue properties would be better, in part because their lower elastic modulus means that the range of tendon tensions induced by live load on the deck would be less.

Strengthening of Prestressed Concrete Bridge Decks by Added Prestressing

22. This technique has been used on several strengthening projects in the past. However, to the author's knowledge, none were undertaken without traffic disruption.

23. An early example was the Newmarket viaduct, built in New Zealand in the sixties as an in-situ prestressed concrete segmental box girder deck using staged cantilever construction. The deck had to be strengthened with prestressing tendons installed inside the box girders to counter the unexpectedly high differential temperature stresses which developed and which had not been allowed for in the original design.

24. A more recent example was the repair and restoration to full strength of some of the viaducts on the A3 highway in southern Italy (Reference 4). The decks were constructed as a series of simply supported 31.3 m post-tensioned concrete beams supporting and composite with reinforced concrete deck slabs. After some 25 years service the decks required extensive repair due to the corrosion damage caused by road de-icing salts. It was also found that the prestressed beams had suffered far greater prestressing tendon losses than estimated in the original design. In one case, existing compressive stresses were measured and estimated to be as low as 30% of theoretical design values.

25. Where the existing prestress was found to be 40% of theoretical, or less, the old deck was removed and replaced with a new construction. In the cases of lower prestress losses, between 40% and 90%, it was found to be more economic to restore the strength of the existing beams by adding prestressing. New prestressing cables were incorporated into the beams by enlarging their bottom flanges. The anchorages for the added cables were provided by steel frames or concrete enlargements of the original beam end blocks. Again, it is apparent that the strengthening measures required considerable traffic disruption, although probably as much due to deck slab corrosion repair as beam strengthening.

26. Strengthening of existing prestressed concrete decks by added prestressing is undoubtedly possible with less, if any, traffic disruption. External cables, used in the manner described previously, offer the best possibilities. Firstly because they require minimum interference to the existing deck and secondly because it is easier to incline the cables to provide any additional shear resistance required to counter the increased deck loading. For simply supported spans, cable inclination also reduces the possibility of undesirable added bending tension in the top of the deck near the supports.

27. The type of existing deck construction is also important when considering strengthening by added prestressing. An existing prestressed concrete continuous box girder with ample internal void dimensions offers the best possibility. It allows anchorage, deflection and prestressing of the strengthening cables within the confines of the girder with minimum, if any, disruption to traffic. A beam and slab deck with generous beam spacing offers similar advantages. On the other hand, decks using contiguous precast pre- or post-tensioned beams present difficulties because of lack of access between beams for installing new cables.

28. Once again, it is apparent that the external prestressing proposed would be best effected by the use of high strength fibre cables (Reference 5).

Deck Bending Relief Strengthening by Contra-Loading

29. Professor Klaiber of Iowa State University, who has undertaken much research over the past decade into strengthening steel/concrete composite decks, recently released details of a more direct use of prestressing for deck strengthening (Reference 1).

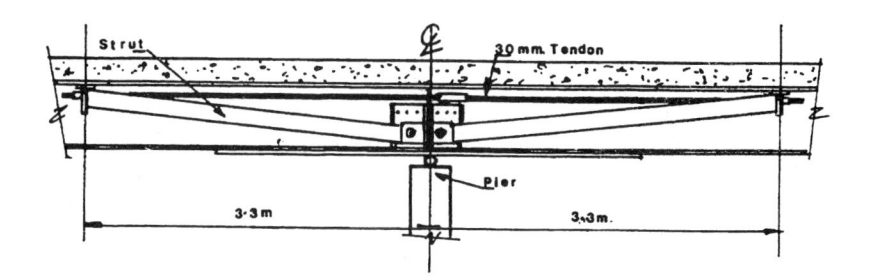

Figure 7
Deck Contra-Loading by Prestressed Truss

PART 2: CASE STUDIES

30. A three span continuous steel/concrete composite bridge in north central Iowa was recently strengthened partially by prestressing bars used in much the same position as the Rakewood Viaduct strengthening, described earlier. The bars were anchored within the girder spans but, in this case, placed on top of the bottom flange. However, the remainder of the strengthening was undertaken by special superimposed trusses laid over the piers either side of the web and within its depth, Figure 7. By tensioning the tendon forming the upper chord of the truss, upward contra-loads are produced at the truss ends to add further direct load relief to the deck self weight bending and shear. The attraction of this new truss form of strengthening prestressing is that no prestress is directed into the deck itself, as required by the earlier methods described. This eliminates the high local stresses associated with prestressing anchors, which have to be superimposed on to the original service load stresses, generally requiring the addition of local web stiffeners.

DECK STEEL/CONCRETE COMPOSITE BEAM FATIGUE LIFE EXTENSION

31. Existing steel/concrete composite bridge decks often require strengthening or fatigue life enhancement of the shear connection between the concrete deck slab and steel girders. This can be undertaken by installing additional new shear connectors. The innovation comes with how to do this without traffic interference.

Figure 8
Welded Stud Shear Connectors on Original DLR Viaduct

32. The existing new viaduct decks of the London Docklands Light Railway, completed in 1987, are generally of continuous composite construction with an in-situ reinforced concrete deck slab supported by and composite with twin steel universal or plate girders. The original design of the decks to BS 5400 established that fatigue considerations were a critical factor, particularly in the deck welded shear connectors, Figure 8. The subsequent decision to build the enormous Canary Wharf Development straddling the railway, Figure 9, resulted in large increases in the weight and frequency of trains after completion, with considerable reductions in fatigue life, in some cases as much as 75%. Strengthening measures to restore the fatigue life back to the originally designed 120 years were required. Additional shear connectors installed between the original 19 mm welded stud connectors would relieve the loads on these connectors sufficiently to accomplish this.

Figure 9
DLR and Canary Wharf Development

PART 2: CASE STUDIES

33. The provision of new shear connectors by drilling-in from under the top flange of the steel girders was examined. Several types of connectors were considered, including 20 mm diameter spring steel pin fasteners. These offered the advantage of a readily achieved force fit into the hole drilled through the steel flange and lower section of the concrete deck slab with no requirement for grouting, gluing or welding.

34. Strength and fatigue testing were carried out on push-out samples by the Welding Institute at Cambridge. 'Spirol' spring pin shear connectors were shown to have superior strength and fatigue properties to the 19 mm studs. The pins obtain their force fit by jacking the lead-in chamfer into drilled holes with slightly smaller diameters, Figure 10. The spring mechanism is generated by the compression of a 2¼ turn spirally coiled strip steel. Good interface shear connection is established, with a degree of pullout resistance afforded by the spring loaded friction between the pin and the hole faced. In the event, the pins were successfully installed with no interruption to the train services, Figure 11 (Reference 6).

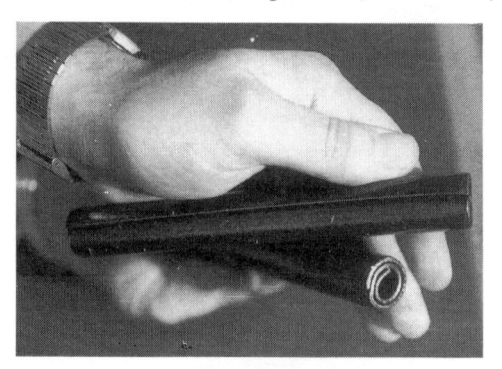

Figure 10
Spring-Pin 'Spirol' Shear Connectors

Figure 11
Jacking 'Spirol' Pins into Place

35. A recent proposal is to use 'Spirol' spring pin shear connectors to improve the fatigue life of the light rail crossing of the Fraser River in Vancouver, Canada.

BRIDGE SUBSTRUCTURE STRENGTHENING
General

36. The UK programme of bridge strengthening to cater for 40t lorries, referred to earlier, will considerably add to the number of existing simply supported multi-span viaducts requiring substructure strengthening. Some of the integrity assessments will indicate that the supporting piers are understrength due to increases in the traction and braking loading since the original design, possibly accompanied by structural integrity damage generated by road salt, carbonation or ASR.

37. In general, any repairs and strengthening procedures are less disruptive to carried road traffic than deck repair and strengthening because the work takes place under the deck. Nevertheless, any cutting away and subsequent strengthening of the piers will prudently require some disruption to the carried traffic. It is also possible that such operations will disrupt heavily trafficked roads, railways or waterways crossed by the bridge. With the costs of disruption so high, every effort should be made by the bridge designer to strengthen piers and abutments with little or no disruption.

38. The shock transmission units described below offer such possibilities to the designer and the sections of the paper which follow demonstrate how substructure strengthening can be achieved with retrofit units by load sharing amongst the piers of a viaduct or between adjacent viaducts.

Shock Transmission Units

39. A shock transmission unit, or STU, is a special mechanism joining separate structures or separate elements of structures. The unique property of the STU is that it permits slow long-term movements between the structures with negligible resistance, yet is capable of acting temporarily as a rigid link between the structures, transmitting short duration shock or impact forces (Reference 7).

40. This means that a series of structures or structural elements subject to long-term separation movements, such as temperature expansions or contractions, can be linked together with STUs to beneficially share amongst them short duration dynamic loads applied to any one of the structures or structural elements. With, say, five equal stiffness structures, a dynamic load applied randomly to any one structure will be resisted equally by five structures. This leads to a reduction of dynamic loading on each structure support system of 80%. In general, it is not possible to link the structures permanently together with simple rigid structural elements

PART 2: CASE STUDIES

because large extra forces would be generated by the restraint to long-term separation movements.

41. STUs have found a number of important applications in all types of engineering over the years. Hitherto the STU has been a relatively complex oil/gas filled device with a high first cost and a continuing need for regular and expensive maintenance and adjustment. This tended to limit the number of applications of this very useful mechanism.

42. In the seventies a new type of STU was invented. A unique chemical compound of boron filled di-methyl siloxane, known as silicone putty, had been developed in America in the sixties and used in the Space Exploration programme. It offered special thixotropic properties, readily deforming under slowly applied pressure but acting as a rigid body under impact. John Chaffe and the late Reg Mander, both engineers with the then UK Ministry of Transport, developed and patented a shock transmission unit using the new silicone putty, initially targeted for use in bridges. Involving only one moving part, it offered structural designers a low cost, robust and minimum maintenance STU for the first time.

43. The typical 50 tonne capacity unit illustrated in Figure 12 uses a steel cylinder containing a loose fitting piston fixed to a transmission rod, the void around the rod being filled with the silicone putty. The unit is attached to the two separate structures or structural elements by fixing eyes located on the cylinder at one end and the transmission rod at the other. The transmission rod passes through the entire length of the cylinder so that the volume of the cylinder and the silicone putty filler remains constant at all piston positions. Under slow movement between the structures the putty is squeezed through the small hole through or the clearance gap around the piston and displaced from one end of the cylinder to the other, generating only small 'drag' forces between the structures.

Figure 12
50 Tonne STU

44. When a short duration impact is applied to one structure, there is negligible movement of the piston, and the impact tensile or compressive force is passed along the transmission rod/piston head/silicone putty/cylinder load path to the second structure. The rating of the unit defines the maximum impact force which can be transmitted, whilst the length of the transmission rod can be varied to suit the expected long term axial movements between the fixing eyes attached to the separate structures. 25-50 tonne units offer particularly compact, lightweight and economic STUs and if larger impact forces have to be catered for, it usually is best to provide groups of the smaller units. It should be noted that the new STU has been designed primarily to function in a horizontal position. However, it can be readily adapted for vertical movement and impact transmission by incorporating an internal spring to return the piston to the neutral position. For all practical purposes the unique thixotropic properties of the silicone putty do not vary significantly through a wide temperature range. Thus the new STU can be relied upon to perform consistently under most climatic conditions.

45. Existing multi-span bridges or viaducts are often made up of a series of simply supported rather than continuous spans. The reasons for this choice may range from an expectation of future large settlements in poor ground or areas of mining subsidence to the economic standardised use of beam units. Whatever the virtues of the simply supported spans for the deck, the associated substructure will usually possess unused design capacity which can be regained for strengthening purposes by adding STUs.

46. A typical simply supported four span/deck is shown in Figure 13. The piers under each simply supported span carry the usual fixed bearings for one span alongside free bearings for the adjacent span. This means of course that the design longitudinal traction and braking forces must be individually applied to each deck span throughout the viaduct. Main resistance is offered by the pier or bankseat carrying the fixed bearings of that particular span, with generally a small additional resistance from friction generated mainly at the free bearings carrying that span, located over the next pier. However, Clause 5.14.2.2 of BS 5400 states that free bearing frictions "are not applicable when calculating stabilizing forces against externally applied loads".

47. This means that a substructure of this type, with (say) equal stiffness bankseats and piers, has a total designed capacity of four times the required deck design traction and braking longitudinal loads. However, by placing four new STUs at bearing level on the existing free bankseat and the piers, as shown in the lower half of Figure 13, it is possible to share out the traction and braking load acting anywhere on the deck amongst all three piers and both bankseats. This in turn means that the bridge strengthened in this manner can resist up to four times the original longitudinal traction and braking forces. Alternatively, it can suffer a significant loss of

PART 2: CASE STUDIES

substructure structural integrity due to corrosion under the original loading without overstress. This simple addition of STUs offers traffic disruption freedom to strengthened and/or damaged substructure retrofitting.

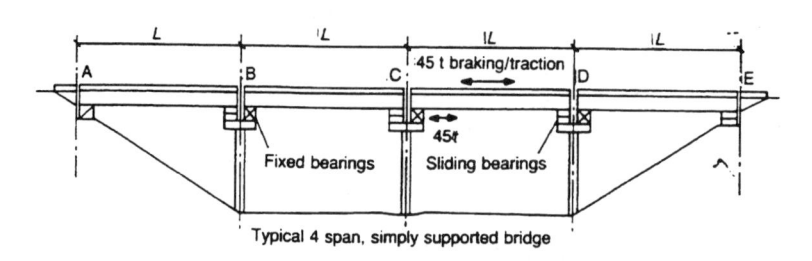

Typical 4 span, simply supported bridge

45 t traction/braking in span	Horizontal load on support: t				
	A	B	C	D	E
AB	45	—	—	—	—
BC	—	45	—	—	—
CD	—	—	45	—	—
DE	—	—	—	45	—

Total support horizontal design capacity required for traction/braking = 45 t + 45 t + 45 t + 45 t
= 180 t

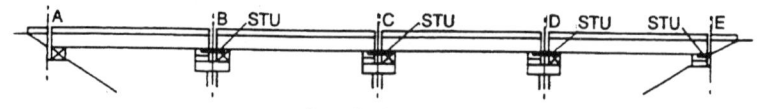

Addition of 4 STUs
(for simplicity assume equal stiffnesses at all supports)

45 t traction/braking in span	Horizontal load on support: t				
	A	B	C	D	E
AB	9	9	9	9	9
BC	9	9	9	9	
CD	9	9	9	9	9
DE	9	9	9	9	9

Total support horizontal design capacity required for traction/braking = 9 t + 9 t + 9 t + 9 t + 9 t
= 45 t

**Figure 13
Using STUs on a Multi-Span Simply Supported Bridge**

**Figure 14
STUs as Part of M5/M6 Midlands Links
Crosshead Replacement**

48. Figure 14 shows STUs used recently in a different load sharing manner for the complex replacement of a salt-corroded reinforced concrete crosshead supporting a section of the heavily trafficked deck of the M5/M6 Midland Links Viaducts in Birmingham. The removal of the corroded crosshead included twin shear walls which served to not only support the deck edges either side of the deck expansion joint but also to transmit deck traction and braking loads down to the crosshead and pier. The new steel trimmer beams visible took over the slab edge support duty whilst traction and braking loads were transmitted through the strut-tie STU links fixed to the decks either side of the expansion joint and thence to the shear walls, crossheads and supporting piers either side of the crosshead replacement.

STUs for Load Sharing Between Viaducts

49. The rapid development of London's Canary Wharf and the requirement to increase the originally planned capacity of the Docklands Light Railway eightfold has been described earlier. The capacity increase is partially effected by doubling up the existing two car trains shown in Figure 9. The trains are automatic, and braking and traction commands are computer controlled. The horizontal forces generated are resisted by the piers forming the substructure.

50. A typical viaduct is continuous over seven spans between deck expansion joints. Train traction and braking horizontal loading, together with any wind loading, is shared among the slender reinforced concrete piers, which generally support the deck by means of rubber bearings.

PART 2: CASE STUDIES

51. Doubling the length of the trains during 1991 meant considerable increases in the horizontal loads arising from the traction and braking efforts. The existing substructures were not designed for this increased loading and original proposals involved extensive strengthening of bearings, piers and foundations, which could have been a very costly and disruptive procedure whilst attempting to maintain an electrically operated driverless train service.

52. The load sharing idea was therefore adapted for the DLR substructure strengthening and the simple introduction of STUs at the deck joints between viaducts, Figures 15 and 16, avoided the costly and disruptive strengthening procedures originally proposed. Some of the increased traction and braking affecting a traversed seven span viaduct are transmitted by impact through the STUs into the adjacent viaducts. The load sharing is sufficient to keep the horizontal loading on the bearings, piers and foundations of each multi-span viaduct within the original design limits.

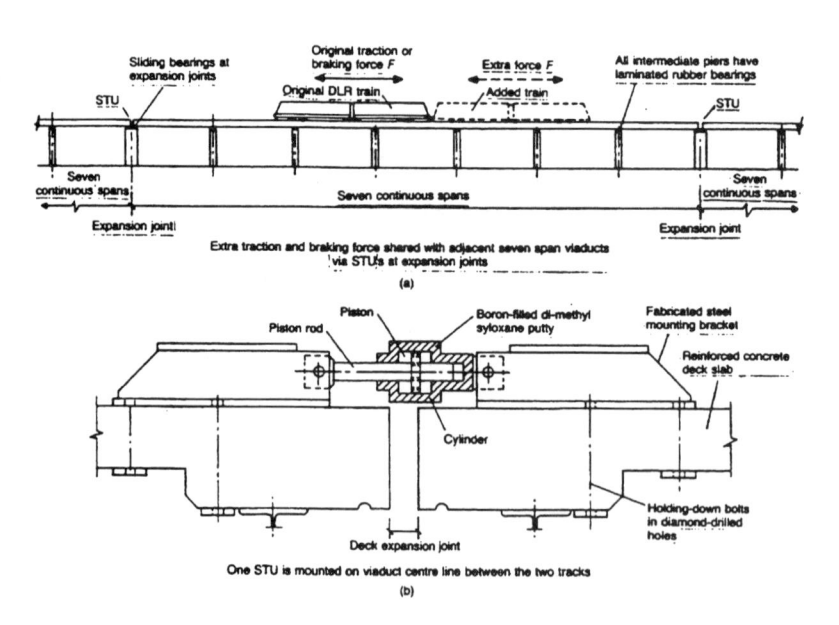

Figure 15
Horizontal Load Sharing Between DLR Viaducts

Figure 16
DLR STUs

REFERENCES
1. KLAIBER, F W, DUNKER, K F, WIPF, T J AND FANOUS, F S. "Strengthening of Existing Bridges by Post-Tensioning", Symposium on Practical Solutions for Bridge Strengthening and Rehabilitation, National Science Foundation and Iowa State University, Ames, Iowa, April 1993.
2. HARRIS, P. "Bridge within a Bridge", Concrete Magazine, London, January 1980.
3. PRITCHARD, B P. "Strengthening of the M62 Rakewood Viaduct", Construction Marketing Symposium on Strengthening and Repair of Bridges, Leamington Spa, June 1988.
4. PETRANGELI, M P. "Inspection and Repair of Some Highway Bridges in Italy", First International Conference on Bridge Management, University of Surrey, March 1990.
5. WOLFF, R AND MIESSELER, H J. "Prestressing with Fibre Composite Materials and Monitoring of Bridges with Sensors", First International Conference on Bridge Management, University of Surrey March 1990.
6. PILGRIM, D AND PRITCHARD, B P. "Docklands Light Railway and Subsequent Upgrading, Design and Construction of Bridges and Viaducts", Proc Inst Civil Engineers, August 1990.
7. PRITCHARD, B P. "Shock Transmission Units for Bridge Design, Construction and Strengthening", Construction Repair, October 1989.

Plate bonding: a user's guide note

A. M. HENWOOD, Group Engineer, W. S. Atkins Consultants Ltd, and K. J. O'CONNELL, Project Engineer, Costain Building and Civil Engineering Ltd

SYNOPSIS. The use of externally bonded steel plates to strengthen concrete highway structures has been the subject of considerable interest in recent years, but the number of structures which have been strengthened by this technique remains relatively small. One of the largest examples to date was at Bolney Flyover on the A23 London to Brighton Trunk Road where 676 plates were applied to strengthen the deck. This paper draws on the experience gained at Bolney Flyover, approximately one year after the plates were installed, in order to highlight the problem areas for future designers and contractors and then reviews the current status of plate bonding as a strengthening technique.

INTRODUCTION

1. The strengthening of Bolney Flyover has been described in detail elsewhere (Reference 12) and so only a brief description of the works are presented here:

2. Bolney Flyover is a four span structure carrying the A23 London To Brighton Trunk Road over the A272 in West Sussex. It was constructed between 1970 and 1972, with sufficient width to accommodate the widening of the A23 dual carriageway from two to three lanes in each direction. Each carriageway is carried on separate in-situ reinforced concrete decks, 762 mm thick with 457 mm diameter circular voids. The decks are supported on reinforced concrete bank seats and columns with driven steel piled foundations. The bridge is on a skew of approximately 26° and each of the four skew spans is 14 m long (See Figure 1).

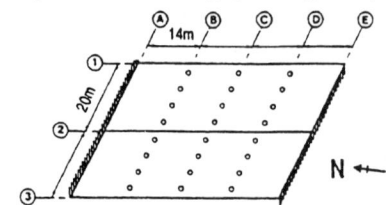

Figure 1 - Plan View of Deck Showing Supports

3. As part of the A23 Improvement Scheme, the bridge was assessed and found to be deficient. The use of externally bonded steel plates to strengthen the deck was considered to be the most appropriate solution. In total 676 plates, 360 mm wide and 6 mm thick, were applied in up to three layers with 6720 fixings, requiring careful attention to procedures and details. The work was commenced in July 1992 and successfully completed in December 1992. Areas that required careful consideration are discussed under the following headings of Assessment, Design, Specification and Construction:

ASSESSMENT

4. The bridge was to be incorporated in the improvement works for the A23, which involved widening to three lanes in each direction at this location, and it therefore required assessment to the current Department of Transport Standard for Highway Loading, BD 37/88.

5. An Assessment Inspection was carried out in September 1989 together with a Special Inspection involving site testing, sampling and laboratory testing. The inspection identified the presence of longitudinal cracks on the deck soffit, directly beneath each carriageway, measuring up to 0.55 mm in width (see Figure 2). The deck was otherwise in good condition with no indication of significant deterioration.

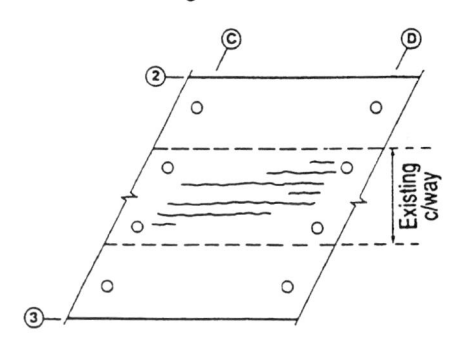

Figure 2 - Crack Pattern

6. A finite element analysis of the deck (using both uncracked and cracked section properties) under the highway loading of BD 37/88 for the widened carriageway arrangement established that the transverse distribution reinforcement did not comply with either serviceability or ultimate limit state requirements for flexure. However, it did satisfy current shrinkage and early thermal cracking requirements. Under existing loading conditions, the assessed crack width was 0.3 mm for permanent loads and 0.6 mm with live loading.

7. This confirmed that the bridge exceeded the serviceability limits for cracking in the transverse direction, and indeed may have also exceeded its ultimate strength depending on the actual level of load experienced.

8. Notwithstanding these findings, a yield line analysis demonstrated that, although these inadequacies existed in the transverse direction, the bridge had adequate overall strength to carry the required loads without collapsing. However, in order to prevent the opening of new cracks when the carriageway was widened and in order to restore the transverse distribution qualities of the deck, the decision was taken to strengthen the deck and the use of externally bonded steel plates across its full width was considered appropriate.

9. Minor serviceability inadequacies were also predicted in the top of the deck at the transverse support sections (0.4 mm crack widths under live loading), but these were not considered significant as a new waterproofing membrane was to be provided. The assessed crack widths were accepted by the Department as a departure from standard. Although it would be difficult to carry out future inspection of a plate bonding system beneath the surfacing, with adequate protection there is no reason why plates should not be applied to top surfaces of bridge structures.

10. The decision to use externally bonded steel plates as the means of strengthening the deck was based on a number of factors :

(a) the capacity of the existing foundations did not facilitate increasing the depth of the deck;
(b) the available headroom was sufficient to accommodate the plates;
(c) the disruption of traffic would be minimised;
(d) the soffit is easily accessible for future inspection;
(e) plate bonding was estimated to be cheaper than demolition and reconstruction of the deck;
(f) there was increasing confidence in this method of strengthening after seventeen years experience in the United Kingdom.

DESIGN
Basic Principles

11. The design of the system was based on the Department of Transport's Draft Advice Note BA 30/89 "Strengthening of Concrete Highway Structures using Externally Bonded Steel Plates". A draft revision of this Advice Note, BA 30/92, has subsequently been prepared.

12. The principle is that the adhesive enables the steel plate to act compositely with the existing section and carry transient loads. The permanent load is already carried by the existing section on its own, since no temporary supports are involved. This avoids any consideration of creep in the adhesive.

13. Elastic stresses were calculated in the original section due to the permanent loads and then in the strengthened section due to the live loads. The principle of superposition was applied to determine the final stresses which were checked against serviceability limits. The ultimate moment of resistance of the strengthened section was also calculated and a check made to ensure that the section was not "over-reinforced".

Plate Design

14. The plates were 6 mm thick, 360 mm wide and the rows were spaced at 1000 mm centres (see Figure 3). In the end spans two layers of plates were required to achieve the necessary structural capacity. Sufficient additional plates were provided to satisfy both serviceability and ultimate limit state bending moments from the original uncracked analysis of the deck.

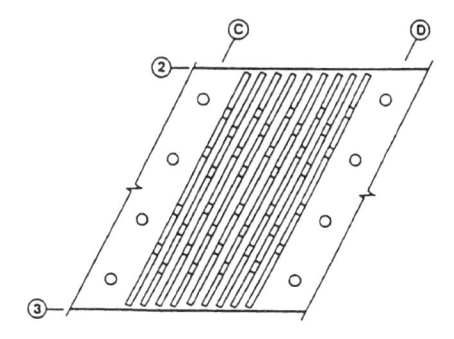

Figure 3 - Plate Arrangement

Bolt Design

15. Research has established that concentrations of both shear and peeling stresses occur at the ends of the plate and for this reason BA 30/89 required the provision of anchorage bolts to resist three times the ultimate limit state longitudinal shear stress.

16. No guidance is given on the design of these bolts and therefore reference was made to manufacturer's literature for information on tensile loads, shear loads, embedment depths and tightening torques. In particular, there is no advice available about required hole diameters in the plate and so, in order to ensure transfer of shear load to the bolts, the annulus around the bolt shaft was filled with epoxy resin adhesive. In addition the adhesive bonding the plate to the soffit was designed to be capable of resisting three times the ultimate limit state longitudinal shear stress.

17. The provision of bolts in plate bonding schemes typically represents 25% of the overall cost, but there is uncertainty about the behaviour of the

bolt/plate connection. This indicates a need for further research in this area concentrating on exactly how much load is transferred to the bolts.

18. Plates should not generally extend into areas of compression as plate buckling may occur causing tensile peeling stresses in the adhesive. Where this cannot be avoided, BA 30/89 required the provision of bolts at a spacing not exceeding 32 times the thickness of the plate or 300 mm whichever is the lesser. At Bolney Flyover, certain combinations of load may cause the plates in the internal spans to go into compression and therefore bolts at a spacing of 192 mm were required. However, this was considered too close for practical purposes.

19. An elastic critical buckling analysis was carried out, assuming the plate to be a two-pinned strut and an increased spacing of 380 mm was proposed. This approach was subsequently accepted by the Department during the Contract, subject to a maximum spacing of 300 mm, and this has now been included in BA 30/92 for plates with low compressive stresses. This resulted in a significant reduction in the total number of fixings originally required from 9360 to 6720, with a corresponding reduction in cost.

Bolt Detailing

20. When detailing the bolt positions, careful consideration of the existing reinforcement arrangement is necessary together with an assessment of the ease of locating this reinforcement. The skew arrangement of reinforcement at Bolney Flyover caused problems in determining an appropriate spacing for the bolts and made the accurate location of the reinforcement difficult (See Figure 4). Orthogonal arrangements are more easily accommodated.

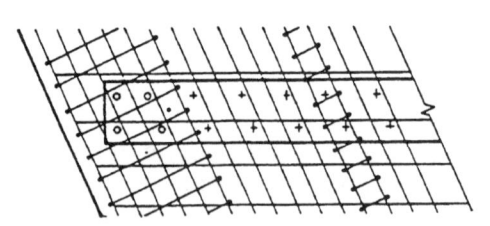

Figure 4 - Plan showing existing reinforcement

21. It is therefore prudent to design spare capacity into the bolt groups in case of clashes with existing reinforcement, and in some circumstances section analysis assuming some loss of existing reinforcement due to drilling may also be appropriate. At Bolney Flyover no severing of main longitudinal reinforcement was permitted but in certain locations the

severing of transverse reinforcement was tolerable. Generally, clashes with existing reinforcement (which occurred at approximately 10% of the bolt locations) were dealt with by reduced embedment depths, taking advantage of designed spare capacity.

22. Attention must be paid to maximum and minimum bolt spacing and edge distances, bearing in mind the possible reduction in tensile and shear capacity of the bolts if these distances are too close. All bolt positions should have sufficient tolerance to allow for repositioning in order to avoid existing reinforcement.

SPECIFICATION
Removal of Traffic

23. At Bolney Flyover, uncertainty about the effect of traffic vibration on the curing of the adhesive led to the decision not to permit traffic on the deck during the plate bonding operation. Since each A23 carriageway is carried on separate decks the traffic management implications of this decision were not unduly onerous, but at other locations this may not be the case. The carriageway arrangement on the A272 was also beneficial at this location, with two spans available for traffic diversion whilst access scaffolding to the deck soffit was in position.

24. There are examples of bridges which have been successfully plated whilst still remaining open to traffic, but it is believed that research has indicated a possible loss in strength of up to 30% if the adhesive is subject to vibration during curing. Clearly, further research in this area is desirable.

Environment

25. The specification of an enclosed (but ventilated) controlled environment using access scaffolding and screening should be included. The temperature within the area should be monitored and dry air heating provided, if necessary, to maintain the temperature above 5°C, this being the temperature below which most epoxy resins will stop curing. No plate bonding should take place if either ambient or substrate surface temperature is below 5°C. The specification of maximum temperatures may also be appropriate in order to control reduction in pot life of the adhesive.

26. Preparation of concrete surfaces by grit blasting should be isolated from plate bonding operations.

27. All operatives should be instructed in the necessary precautions associated with the safe use of plate bonding materials. Particular reference should be made to the harmful vapours produced by some solvents used with adhesives. Any waste materials retained on site should be stored in a safe and secure manner and disposed of with due regard to current regulations.

Steel Plates

28. Once the covermeter survey is complete, the final positioning of the plates and bolt holes to avoid the existing reinforcement is best determined by marking on the soffit itself. Fabrication drawings can then be prepared and submitted to the Engineer for agreement. The use of templates may be appropriate where reinforcement is orthogonal and bolt arrangements are repetitive.

29. When more than one layer of plates is to be applied, care should be taken to ensure that the holes in the different layers match, taking due account of any fabrication tolerance in plate length. The specification of match drilling (one plate on top of another) may be appropriate in these circumstances.

Adhesive

30. Compatibility of adhesive and primer should be specified and this should be backed up by tests on bond strength and durability. It may also be appropriate to specify compatibility of the adhesive with any repair mortars that may be used during preparation of the concrete substrate and in this respect cooperation between suppliers of chosen products is beneficial.

Corrosion Protection

31. Although not mentioned in BA 30/89, the application of primer to the bonded face of the plate is now recommended to assist durability (Reference 4). This is still not clearly specified in BA 30/92 although it is implied within the section on surface preparation.

32. The application of the primer has a significant effect on the strength of the adhesive joint, the shear strength being influenced by both the age and thickness of the primer. There is little research evidence relating the structural capabilities of the primer to the performance of the joint. For such a key element of the system this is a significant shortfall in knowledge.

33. In order to ensure total compatibility at Bolney Flyover the paints came from the same manufacturer as the adhesive. The first coat was a two part micaceous iron oxide epoxy primer and the second and third coats were high build PVC acrylic undercoat and finish. The minimum total dry film thickness was 200 microns.

Testing

34. The following tests were agreed for Bolney Flyover:
    (a) <u>Flexural Modulus of Elasticity</u>. The main purpose of this test was to ensure consistency of mix and therefore one sample was required each time a mix was made. If, for a given combination of batch numbers, initial results prove favourable, the frequency of testing for that combination may be gradually relaxed to a minimum of one per

day. However, every permutation of batch numbers used should be tested.
(b) <u>Bond to Steel Substrate</u>. For this test five double lap shear tests should be carried out on each batch of adhesive delivered to site (as identified by a batch number on the tins). Resin and hardener tins delivered together should not be separated, and again every permutation of resin and hardener used should be tested. The primer film thickness on the test specimens should be the same as on the main plates.
(c) <u>Bond to Concrete Substrate</u>. A minimum of six pull-off tests using the contract adhesive should be carried out on selected areas of the soffit previously prepared for plate bonding.

35. All test specimens should be assembled and cured on site using samples of adhesive mixed for application to the main plates. The record sheets should demonstrate that due attention is being paid to the shelf life of the tins.

36. At present there is no guidance on the minimum age of the primer or the adhesive required at the time of the tests. This can cause confusion in the interpretation of results, especially when these fall short of specified levels. At Bolney Flyover, almost without exception, the failure paths in the double lap shear test specimens were all through the primer layer, emphasising the importance of understanding the effect of the primer on the structural capability of the joint. It is important to examine the double lap shear test specimens carefully in cases where results fall below expected values in order to correctly identify the cause.

37. It is unfortunate that the results from Tests a) and b) will not generally be available until the plate is installed. However, if results for a particular mix prove unacceptable, removal of the plate is possible using a wedge cleavage type action. Notwithstanding this, the specification of test response times is recommended.

<u>Trial</u>

38. It is advisable to carry out a trial to demonstrate the suitability of the installation proposals, which may be unique for any particular structure.

39. If possible it is considered advantageous to carry out the plate bonding trial on the structure itself rather than on a specially constructed panel, in order to fully simulate the actual conditions and reduce cost.

40. Initially the erection procedure can be examined by using a debonding layer of tape applied to the soffit. Once a satisfactory performance has been demonstrated, the tape can be removed, the trial plate erected and the fixings installed.

PART 2: STRENGTHENING TECHNIQUES

41. It may be useful to specify the inclusion of a deliberate void in the adhesive to be used as a reference for audible hammer testing for delamination.

CONSTRUCTION
Resources
42. A consistent installation team was maintained throughout the operation under an autonomous management team.

43. In view of the unusual nature of the work, the need to respond to unforeseen difficulties and the requirement to keep detailed records, it was considered necessary to maintain a high level of supervision throughout the operation and the RE assigned one engineer exclusively to the works at Bolney Flyover.

Covermeter Survey
44. The need for the holes to avoid the existing reinforcement required accurate results from the covermeter survey. To ensure this, several covermeters were investigated and tested. The Protovale CM5 was eventually chosen as the most versatile and reliable for locating the plan position of the reinforcement. A test slab was cast with the same steel arrangement as the existing deck to assess the accuracy of the readings being obtained at the bridge soffit. The main longitudinal steel and shear links, both of which were close to the surface, were accurately located but the identification of the deeper transverse bars in the second layer was more difficult.

45. As already mentioned, in the event of clashes with existing reinforcement, various different options were available depending on the type of hole and location, but these were essentially based on the spare capacity in the specified bolts, which meant reduced embedment depths could be tolerated.

Crack Injection
46. All cracks wider than 0.2 mm were injected using a low viscosity two part epoxy resin system (283 m in total). It is recommended that cores are taken to check the success of this operation.

Plate Application
47. The specification assumed an erection sequence in which the adhesive is applied to both the plate and the concrete prior to erection, and the bolts are fixed into the concrete after the application of the plates.

48. Application of the adhesive to both surfaces was considered to be the preferred method since it ensured proper wetting of both concrete and steel substrates. Profiling of the adhesive cross-section across the plate, such that it was thicker in the middle than at the edges, ensured that when the plate

was pushed into contact with the soffit the adhesive was pushed from the centre to the edges, encouraging mixing at the adhesive interface and reducing the risk of air entrapment. It was necessary to pay careful attention to the thickness of adhesive at the plate ends where greater loss of material occurs during plate application.

49. Since so many bolts were involved, it was not considered feasible to fix the bolts into the soffit prior to erection of the plates and then achieve perfect matching of the plate holes with the bolts.

50. Adopting the assumed erection sequence, a simple and versatile lifting and support system was devised which accommodated all the variations in plate arrangement and kept manpower to a minimum. The plate was pressurised against the soffit by wedging off a reaction beam which was lifted and locked into position by a system of ropes and pulleys.

51. A quick and efficient system was very important to ensure that the time between mixing the first batch of adhesive and the final positioning of the plate against the soffit was kept within the pot life of the material (typically 40 minutes at 20°C).

Rate of Plate Bonding

52. The application of the externally bonded steel plates was carried out in conjunction with other major items of refurbishment work on the bridge which included replacement of bearings, expansion joints and waterproofing, reconstruction of abutments and deck edges and construction of column protection barriers, all within a major £16m improvement scheme on the A23, with traffic flows on both the A23 and A272 maintained at all times.

53. In total 676 plates were erected over a period of 26 weeks. However, this included the initial slow start whilst teething difficulties were resolved and a 6 week delay due to phasing of other works at the bridge. The site team erected an average of 10 plates per day, but this peaked at 25 plates per day three-quarters of the way through the operation.

PERFORMANCE IN SERVICE

54. Inspection of the structure after approximately one year in service has confirmed satisfactory performance. The adhesive interface has been inspected visually and audibly (listening for indications of delamination by hammer tapping) and shows no signs of deterioration. The corrosion protection system is intact and the sealed cracks have not re-opened

PART 2: STRENGTHENING TECHNIQUES

CURRENT STATUS OF PLATE BONDING
Department of Transport Structures

55. Final issue of the Departmental Advice Note BA 30/92 has been delayed whilst consideration is given to the acceptability of using plate bonding for strengthening at ultimate limit state. There is concern that if a failure in the adhesive bond goes undetected between routine inspections, a bridge may be at risk of collapse. Clearly, since the technique is generally used for ultimate strengthening, the potential use of plate bonding on Department of Transport structures would be limited.

56. However, the problem needs to be kept in perspective since structures strengthened with several rows of plate (where failure of one row is not significant) would be less at risk than those with only a few rows.

Alternative Materials

57. The reason for the lack of confidence in the adhesive bond is due to the possibility of corrosion occurring at the steel/adhesive interface. For this reason steel plates are currently given a coat of primer prior to bonding, but clearly there is a potential future for materials which do not suffer corrosion problems such as fibre reinforced plastics.

58. Fibre reinforced plastics can now be produced with better mechanical and physical properties than steel but research in Switzerland (Reference 6) has identified that the ultimate failure mode of strengthened beams can be sudden and without warning. Notwithstanding this, two bridges in Switzerland have already been strengthened using carbon fibre reinforced plastic plates.

59. Although more expensive than steel, thinner plates are required and the lighter weight is distinctly advantageous on site in what is essentially a labour intensive operation.

CONCLUSIONS

60. The continuing feasibility of using externally bonded steel plates to strengthen bridge decks was demonstrated at Bolney Flyover. Its successful use at other locations is dependent not only on careful attention to procedures and detail but also on careful assessment of the implications of bond failure. Further research is required in the following areas:

- The acceptability of bonding to bridges which remain open to traffic.
- The behaviour of the anchorage bolts.
- The long-term durability performance of the system.
- The effect of the primer on the strength of the adhesive joint.
- The use of alternative materials such as fibre reinforced plastics.

LIST OF BRIDGES STRENGTHENED USING PLATE BONDING

61. This list covers the UK only, is not exhaustive and does not cover all areas.

Date	**Bridge**	**County**
1974	Dedham Bridge	Suffolk
1975	M5 Quinton Interchange	Hereford and Worcester
1977	M20 Swanley Interchange	Kent
1977	A4120 Pen-y-Bont Bridge	Dyfed
1977	Pen y Llyn Isaf	Gwynedd
1982	M1 Brinsworth Road Bridge	Yorkshire
1986	A10 Brandon Creek Bridge	Norfolk
1987	M2 Farthing Corner Footbridge	Kent
1987	M1 Stainsby-Teversal Bridge	Derbyshire
1988	Olive Road Railway Bridge, Hove	East Sussex
1989	A56 Cedars Overbridge	Cheshire
1989	Manor Road Railway Bridge	East Sussex
1990	Finavon Bridge	Forfar, Scotland
1990	Austen Fen Bridge	Lincolnshire
1991	Duckes Marsh Bridge	Devon
1991	Stakeford Bridge	Northumberland
1991	Bures Bridge	Suffolk
1991	Bridgewater Road Bridge, Brent	London
1992	Quidhampton Bridge, Basingstoke	Hampshire
1992	A23 Bolney Flyover	West Sussex
1992	Mythe Bridge, Tewkesbury	Gloucestershire
1993	Great Bridge, Romsey	Hampshire
1993	Duttons Road Bridge, Romsey	Hampshire
1993	Cramlington Subways	Northumberland
1993	Plas Bridge	Gwynedd
1993	Kirkstead Bridge	Lincolnshire
1993	Newbridge, Twyford	Berkshire
1994	Park Hill Road Bridge, Bexley	London
1994	Market Harborough Bridge	Leicestershire
1994	Pont Geinas Bridge	Clwyd
1994	Wisbech Town Bridge	Cambridgeshire

ACKNOWLEDGEMENT

The paper is published with the permission of the Director of the South East Construction Programme Division of the Department of Transport and the authors acknowledge the assistance of all colleagues involved in the strengthening works carried out at Bolney Flyover.

REFERENCES

1 RAITHBY, K D. External strengthening of concrete bridges with bonded steel plates. Supplementary Report 612, Transport and Road Research Laboratory, Crowthorne, 1980.

2 LADNER, M and WEDER, Ch. Concrete structures with bonded external reinforcement. Report Nº 206, EMPA, Dubendorf, 1981.

3 MACDONALD, M D. The flexural performance of 3.5 m concrete beams with various bonded external reinforcements. Supplementary Report 728, Transport and Road Research Laboratory, Crowthorne, 1982.

4 CALDER, A J J. Exposure tests on external reinforced concrete beams - performance after 10 years. Research Report 129, Transport and Road Research Laboratory, Crowthorne, 1988.

5 MAYS, G C and HUTCHINSON, A R. Engineering property requirements for structure adhesives. Proceedings of the Institution of Civil Engineers, Part 2, Volume 85, September 1988, pp 485-501.

6 MEIER,U. and KAISER, H.P. Strengthening of Structures with CFRP Laminates. Proceedings Advanced Composite Materials in Civil Engineering Structures. MT Div / ASCE / Las Vegas, Jan 31, 1991.

7 MAYS, G C and TURNBULL, J D. Strengthening bridges with bonded external reinforcement. 1992 Bridge Symposium, Leamington Spa, May 1992.

8 DEPARTMENT OF TRANSPORT. Departmental Advice Note BA 30/89 (Draft) Strengthening of concrete highway structures using externally bonded plates.

9 DEPARTMENT OF TRANSPORT. Departmental Advice Note BA 30/92 (Draft) Strengthening of concrete highway structures using externally bonded plates.

10 SHAW, M. Strengthening bridges with externally bonded reinforcement. 2nd International Conference on Bridge Management, April 1993.

11 MAYS, G C. The use of external reinforcement in bridge strengthening : Structural requirements of the adhesive. 2nd International Conference on Bridge Management, April 1993.

12 HENWOOD, A M and O'CONNELL, K J. The use of externally bonded steel plates to strengthen Bolney Flyover. 2nd International Conference on Bridge Management, April 1993.

Enhancing influences of compressive membrane action in bridge decks

A. E. LONG, Director, School of the Built Environment, The Queen's University of Belfast, J. KIRKPATRICK, Principal Engineer, Department of the Environment, and G. I. B. RANKIN, Technical Adviser, Technology Centre, The Queen's University of Belfast

INTRODUCTION

1. In the UK the bridge assessment programme is now in full swing and this has highlighted deficiencies in many different types of bridge structures. Some forms of concrete structures have an inherent strength which is ignored by current design codes. In particular it is accepted that the capacity of the slab elements of beam and slab decks is greatly enhanced due to the restraint provided by the beams and diaphragms. This enhancement has been recognised by many bridge authorities worldwide by incorporating it into their national design codes. Whilst BS5400 does not recognise this the current UK assessment codes (ref. 1, 2) for concrete structures do allow compressive membrane action to be included in the assessed capacity of deck slabs. The recognition of compressive membrane action is most important as it can mean the difference between a bridge deck passing or failing the assessment requirements.

COMPRESSIVE MEMBRANE ACTION IN SLABS
Background and concept

2. With the advent of Johansens (ref. 3) yield line theory in the 1940s designers and researchers felt that at long last they had a prediction method for slabs which would provide realistic strength estimates. However, in 1955 tests carried out by Ockleston (ref. 4) on interior panels of the Old Dental Hospital in Johannesburg revealed collapse loads of 3-4 times those predicted by the yield line method. This enhanced capacity was attributed to the development of an internal arching mechanism arising from the restraining effect of the surrounding panels.

3. Where a beam is restrained against longitudinal expansion, the concept of arching action can best be understood by referring to Fig. 1.

PART 2: STRENGTHENING TECHNIQUES

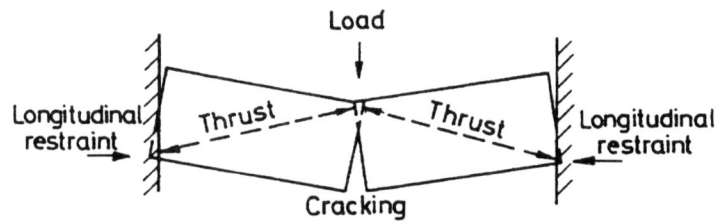

Fig. 1. ARCHING ACTION IN CONCRETE ELEMENT

With the development of tension cracks at mid-span and at the supports the beam tries to expand longitudinally but as it is restrained, corresponding forces are ind.uced which allow it to sustain a substantial load on the basis of the arching thrusts which develop as the deformation increases. Similar actions take place in two way systems where a dome or membrane rather than an arch is generated and this phenomenon is generally referred to as "compressive membrane action" (C.M.A.)

4. The extent of the enhancement provided by compressive membrane action, over and above the flexural strength, depends on the degree of restraint provided by the surrounding structure. A typical load deflection curve with the notional contributions of C.M.A. and flexural action separately identified is given in Fig. 2.

Relevance to bridge deck slabs

5. Tests on model bridge deck slabs in the Civil Engineering Department, Queen's University, Kingston, Canada, in the late 1960s revealed considerable reserves of strength against punching failure (ref. 5). The cause of this enhancement was correctly identified as compressive membrane action and its particular relevance to transient concentrated wheel loads was recognised. Here it is important to note that bridge decks represent one of the first areas to be considered appropriate for the utilisation of these design concepts. This is largely because the major localised loading is transient in nature and hence creep, which may reduce the enhancing effects of C.M.A., is of little importance.

6. A number of small scale model tests was carried out by Hewitt and Batchelor (ref. 6) for the Department of Highways, Ontario and on the basis of the test results a conservative design method was produced. Thus in the 1979 Ontario design standards (ref. 7) for beam and slab bridges, nominal transverse reinforcement (0.3%) only was required to resist concentrated wheel loadings as opposed to the 1.7% normally required on the basis of flexural design. Similar design concepts are now accepted in various states in the U.S.A. and to date no adverse effects have been detected from these reductions in levels of reinforcement.

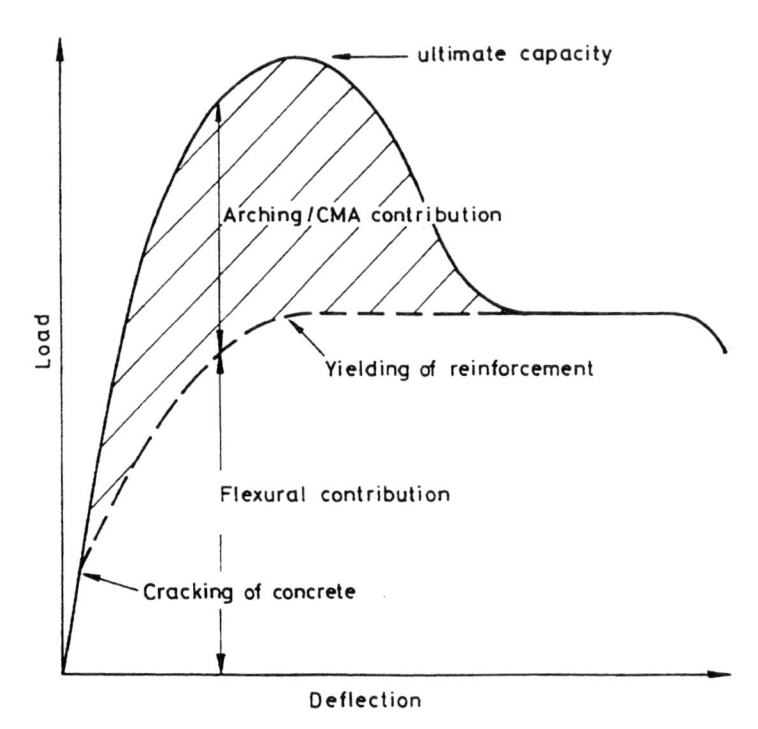

Fig. 2. INTERACTION BETWEEN FLEXURAL & COMPRESSIVE MEMBRANE ACTION

RESEARCH ON COMPRESSIVE MEMBRANE ACTION IN THE U.K.

Validation tests in Northern Ireland

7. In the knowledge of the research carried out in Canada on A.A.S.H.T.O. girder based beam and slab bridge decks it was decided that parallel tests should be carried out on spaced M-beam decks to determine whether similar reductions in transverse reinforcement were possible. This would allow a slightly larger M-beam to be used at a spacing of 1.5m or 2.0m with consequent savings relative to smaller M-beams at 1.0m spacing.

8. In order to establish the strength of the slabs spanning between beams a one-third scale model bridge deck was constructed in the laboratory and tested at Queen's University, Belfast, in the late 1970s. The model was fully representative of a prototype and in particular the end diaphragms and parapet upstand were included to ensure the slab could develop its full potential of in-plane forces. Details of the model deck and the notation for the test panels are given in Fig. 3.

PART 2: STRENGTHENING TECHNIQUES

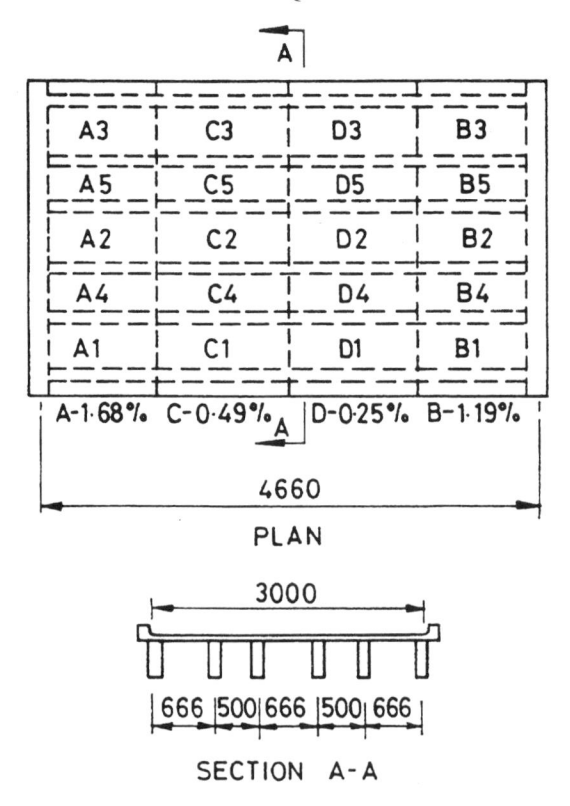

```
                    A
      ┌─────┬─────┬─────┬─────┐
      │ A3  │ C3  │ D3  │ B3  │
      ├─────┼─────┼─────┼─────┤
      │ A5  │ C5  │ D5  │ B5  │
      ├─────┼─────┼─────┼─────┤
      │ A2  │ C2  │ D2  │ B2  │
      ├─────┼─────┼─────┼─────┤
      │ A4  │ C4  │ D4  │ B4  │
      ├─────┼─────┼─────┼─────┤
      │ A1  │ C1  │ D1  │ B1  │
      └─────┴─────┴─────┴─────┘
      A-1·68%  C-0·49%  D-0·25%  B-1·19%
                    A
      |←────────── 4660 ──────────→|
                  PLAN
```

|←────── 3000 ──────→|

SECTION A-A

|666|500|666|500|666|

Fig. 3. DETAILS OF 1/3 SCALE MODEL BRIDGE DECK

9. The design of the prototype slab for the two 112.5kN wheel loads using the equations of Westergaard indicated that steel reinforcement of the order of 1.7% was required under the highway pavement. For test purposes areas of reinforcement equivalent to approximately 1.7%, 1.2%, 0.5% and 0.25% were provided in the model and these were represented by 4 areas of slab approximately 1200mm wide in the span-wise direction and the full width of the deck in the transverse direction. With this banding of the reinforcement, three panels equivalent to 2m spacing and two panels equivalent to 1.5m beam spacing were available providing a total of 20 panels for testing.

10. A properly scaled concrete mix was used, with care being taken to ensure that the maximum aggregate size was consistent with that of the other model dimensions. Furthermore the ratios of tensile strength to compressive strength for the model mix were similar to that for the prototype concrete.

11. Hydraulic jacks were used to load the model with the loading for

the heavily reinforced panels A and B simulating a single 112.5kN wheel load of an HB vehicle. The lightly reinforced panels C and D were subjected to either single or double wheel loading to check the influence of the latter on the behaviour and strength. All panels were incrementally loaded to failure to determine their ultimate capacity and mode of failure.

Model test results

12. The ultimate load capacity of each test panel was the load which caused the loading shoe simulating the wheel load to punch through the slab in the characteristic manner. It was found that there was very little variation in the ultimate load capacity of all the panels even though the transverse reinforcement varied from approximately 0.25% to 1.7%. In comparison with North American (ref. 7) and United Kingdom (ref. 8) design code predictions (Fig. 4) the results of the tests on the

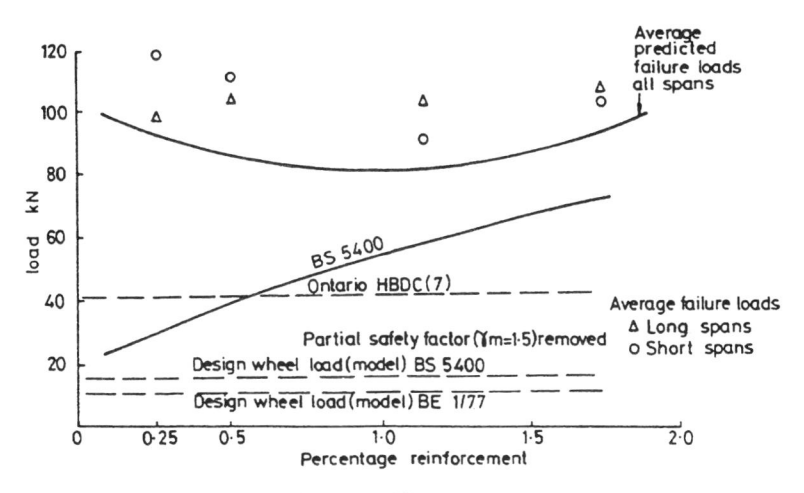

Fig. 4. TEST RESULTS OF MODEL BRIDGE DECK

one-third scale model with the M-beams spaced at up to 2m apart showed considerable enhancement over the design capacity of the standard slab. This enhancement can be attributed to the considerable in-plane restraint that is inherent in bridge slabs. Figure 4 clearly shows that the codes do not give a satisfactory prediction of the punching shear capacity of typical bridge slabs and a more appropriate method which allows for in-plane restraint was therefore developed.

Basis of method of prediction

13. For rigidly restrained bridge slabs, the effect of reinforcement upon the ultimate capacity is small as is evidenced by the results of the

PART 2: STRENGTHENING TECHNIQUES

model test. Thus a method of prediction has been derived to allow for the compressive membrane forces generated within the slab. This method makes use of an effective steel reinforcement ratio, and full details of the method and its derivation are given in ref. 9.

14. As can be seen from Fig. 4 the proposed method of predicting the punching shear strength of reinforced concrete bridge slabs gives good correlation with the results from the one-third scale model.

B.C.A. model tests

15. In 1990 Jackson (ref. 10) reported the results of a series of tests on the slabs of half scale bridge decks. These were carried out in the B.C.A. laboratories and the results obtained for the spaced M-beam configuration confirmed the findings of Kirkpatrick (ref. 9). It was concluded that even though a sophisticated finite element model, which allowed C.M.A. effects to be taken into account, had been utilised, the prediction method of Kirkpatrick was considered to give conservative estimates of the ultimate capacity. In this paper Jackson took account of global as well as local effects and even when the diaphragm was not included in one of the models the measured strength was still higher than that predicted by Kirkpatrick (ref. 9).

Serviceability of deck slabs

16. The ultimate load tests referred to above have indicated that strength is not critical in the design of deck slabs - however, designers also have to satisfy the serviceability limit state requirements.

17. The widths of the cracks induced in the slabs were monitored during the model test in the Queen's University of Belfast and it was found that under service load conditions no cracks resulted. However, as scale effects can affect the accuracy of these measurements full-scale tests (ref. 11) were subsequently carried out on a bridge built by the D.O.E.(N.I.) Roads Service. This bridge incorporated beams at 1.5m and 2m spacing, and the reinforcement varied from 0.25% to 1.7% in the standard 160mm thick deck slab.

18. The tests showed that current crack control formulae are not applicable because of the enhanced performance which results from the development of compressive membrane action. Initial cracking occurred at loads well in excess of the design service loads and even after cracks had been induced by severe overloading it was found that the slabs still satisfied the serviceability limit state requirements.

19. The findings of this research have led to the adoption of a less conservative design approach for M-beam bridge deck slabs by the

Department of the Environment for Northern Ireland (ref. 12). Provided certain restraint conditions are satisfied, the use of a nominal 0.5% reinforcement in the slab is now acceptable.

Slab elements of box girder systems

20. Whilst the slabs of bridges of this type are usually thicker due to their dual function it is conceivable that when assessed on a flexural basis some will be found to be deficient. In this context few, if any, tests to determine the local punching strength of the slabs of box girders have been carried out. However, some useful guidance as to possible reserves of strength may be gleaned from the results of tests by Skates (ref. 13) on cellular concrete structures (Fig. 5). A range of cellular

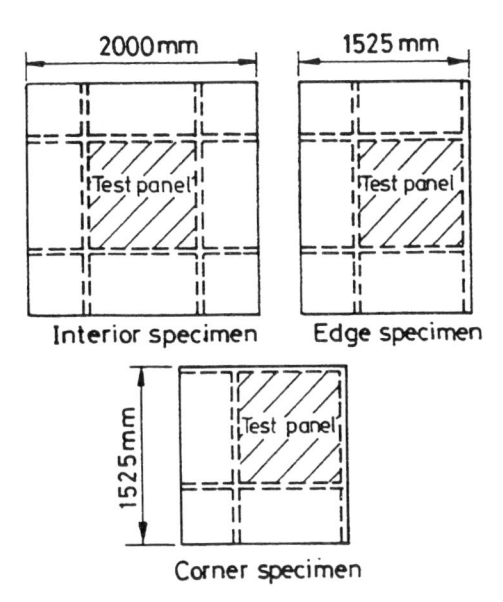

Fig. 5. CELLULAR TEST SPECIMENS

models was tested, one series with concentrated loading and the other series with uniformly distributed loading. The typical results shown in Fig. 6 indicate a marked increase in ultimate load carrying capacity with increasing restraint. In addition it can be seen that enhanced strengths of 3 to 4 times the strength of unrestrained slabs can be achieved for

PART 2: STRENGTHENING TECHNIQUES

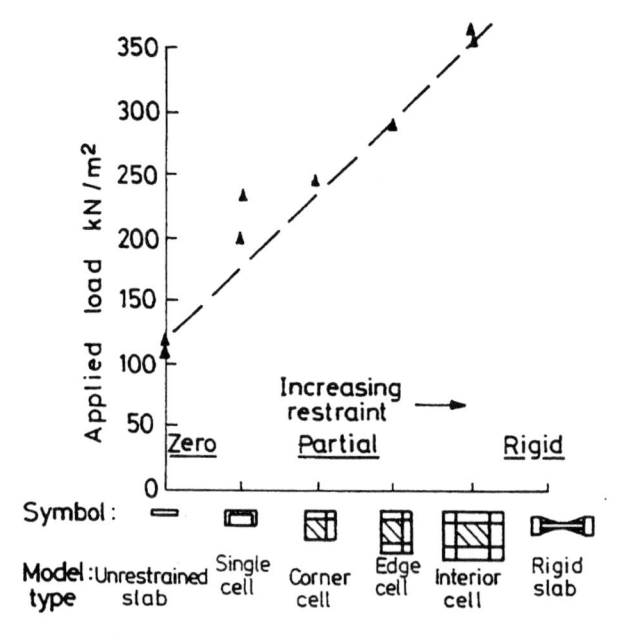

Fig. 6. THE ENHANCED STRENGTH OF CELLULAR STRUCTURES

interior panels and in this regard the results correlate well with the findings of Ockleston. Edge and corner panels also have substantial reserves of strength which would not be taken into account if the slabs had been analysed using yield line theory.

21. Arising out of this series of tests, which were carried out as part of the Marine Technology Directorate's C.O.I.N. (Concrete Offshore in the Nineties) programme, Skates (ref. 14) has developed a design approach based on an extension of Kirkpatrick's work. This could be a useful tool in assessing the strength of this type of structure as it recognises the contribution made by compressive membrane action.

POSSIBLE FUTURE DEVELOPMENTS

22. A common finding of all programmes of tests on slabs which are effectively restrained laterally is that even at very low levels of reinforcement the ultimate strength is more than adequate. Indeed, tests on restrained slabs with no reinforcement have also indicated little or no diminution of strength relative to similarly restrained slabs with reinforcement.

23. Findings of this nature allied to the problem of corrosion of steel in concrete deck slabs have prompted the development in Canada of a novel reinforcing system (ref. 15). In this system the in-plane restraint

to the slab is provided by external steel reinforcement and the control of cracks due to temperature and shrinkage is provided by relatively inexpensive polypropylene fibres. Since these fibres are practically inert to the effects of deicing salts, a deck reinforced by them is not only inexpensive but is likely to be very durable.

24. The feasibility of a number of possible reinforcement arrangements has been assessed via tests on half-scale models of steel beam/concrete slab composite systems. One system which performed well is illustrated in Fig. 7.

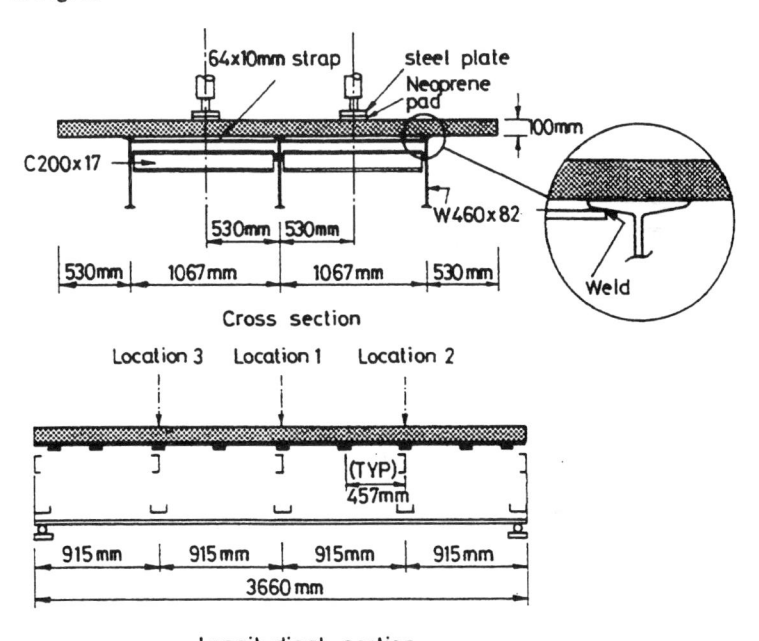

Fig. 7. DETAILS OF THE MODEL TESTED BY MUFTI et al (15)

25. Arising out of these model tests the Canadian authors have enough confidence in the proposed concept to suggest that conservatively designed fibre reinforced concrete deck slabs could now be used in actual bridges.

26. Other alternatives such as the replacement of steel reinforcement by rods of carbon fibres, glass fibres or other non-corroding materials may represent viable alternatives. These could well be cost effective if low percentages of reinforcement were utilised and the potential problems associated with the use of reinforcing materials of low elastic modulus, ie large deformations, are unlikely to arise because of the

enhanced serviceability performance due to the effects of compressive membrane action.

CONCLUSIONS
27. The following conclusions may be drawn from this paper:
1. Compressive membrane action enhances the strength and serviceability of laterally restrained slabs and can be utilised in the assessment of concrete bridge decks.
2. Compressive membrane action concepts have been incorporated into various national design codes and within the U.K. it is taken into account in the N. Ireland regulations.
3. Compressive membrane action may also be utilised to increase the design capacity of the slab elements of box girder bridges.
4. There is considerable potential for the development of highly durable deck slabs, without conventional reinforcement, where the boundary conditions are such that compressive membrane action can develop.

ACKNOWLEDGEMENTS
28. The authors would like to acknowledge the support of the Department of the Environment (N.I.) Roads Service and thank them for their permission to publish this paper.

REFERENCES
1. U.K. Assessment Code. BD44/90. The assessment of concrete highway bridges and structures. The Department of Transport, 1990.
2. U.K. Assessment Code. BD44/90. The assessment of concrete highway bridges and structures (advice note). The Department of Transport, 1990.
3. JOHANSEN K.W. Yield line theory. Translation by Cement and Concrete Association, London, 1962, 181 pages.
4. OCKLESTON, A.J. Load tests on a three storey reinforced concrete building in Johannesburg. The Structural Engineer, October 1955, vol. 33, pp. 304-322.
5. TONG P.Y. and BATCHELOR, B. de V. Compressive membrane enhancement in two way bridge slabs. Cracking, Deflection and the Ultimate Load of Concrete Slab Systems, S.P.-30, A.C.I., Detroit, 1971, pp. 271-286.
6. HEWITT B.E. and BATCHELOR B. de V. Punching shear strength of restrained slabs. Journal of the Structural Division, American Society of Civil Engineers, 101, no. ST.9, September 1975, pp. 1837-1852.
7. Highway bridge design code, Toronto, Ontario, Ontario Ministry of Transportation and Communication, 1979.
8. British Standards Institution, BS5400. Steel, Concrete and Composite Bridges. London, 1978.

9. KIRKPATRICK J., RANKIN G.I.B. and LONG A.E. Strength evaluation of M-beam bridge deck slabs. The Structural Engineer, vol. 62B, no. 2, September 1984, pp. 60-67.

10. JACKSON P.A. The global and local behaviour of bridge deck slabs. The Structural Engineer, vol. 68, no. 6, 20 March 1990, pp. 112-116.

11. KIRKPATRICK J., RANKIN G.I.B. and LONG A.E. The influence of compressive membrane action on the serviceability of beam and slab bridge decks. The Structural Engineer, vol. 64B, no. 1, March 1986, pp. 6-12.

12. Department of the Environment for Northern Ireland. Design of M-beam bridge decks - Amendment No. 3 to Bridge Design Code. N.I. Roads Service Headquarters, 1986, 11.1-11.5.

13. SKATES A.S. et al. Utilizing the effects of compressive membrane action in the design of offshore cellular concrete structures. International conference on concrete in the marine environment, Concrete Society, London, 1986, pp. 79-87.

14. SKATES A.S. Development of a design method for restrained concrete slab systems subject to concentrated and uniform loadings. Ph.D. thesis, Queen's University, Belfast, 1987, 478 pp.

15. MUFTI A.A., JAEGER L.G., BAKHT B. and WEGNER L.D. Experimental investigation of fibre reinforced concrete deck slabs without internal steel reinforcement. Canadian Journal of Civil Engineering, vol. 20, 1993.

Strengthening and modification of bridge supports

M. A. IMAM, Acer Engineering Limited

SYNPOSIS

Bridge supports require strengthening or modification for various reasons. Reappraisal in recent years of vehicle and vessel collision incidents with bridge piers, in many cases leading to disasterous results, is taking strength assessment and modification of bridge supports into a new dimension. This paper sets the scene for strengthening and modification of bridge supports and discusses some techniques which either have or may be employed to achieve the desired results.

INTRODUCTION

1. Bridge supports are perhaps not as glamorous parts of a bridge structure as the decks they support. As a result of this there is more written about the bridge decks than the bridge supports. The most significant modification of bridge supports carried out in recent years is perhaps the strengthening work on the Severn Bridge Towers. However, there is ample material published on this work and therefore no attempt will be made in this paper to devote substantial effort on that project. On the way towards dealing with modification of bridge supports, it is considered pertinent to define BRIDGE SUPPORTS and hence state the scope of this paper.

BRIDGE SUPPORTS

2. Definition of the word "bridge" as found in the dictionary is "a structure that spans and provides a passage over a road, railway, river, or some other obstacle". This definition is somewhat general and is all encompassing. Most engineers however, would be happy to identify any structural element which supports a bridge deck as a bridge support or substructure. Thus for short span bridges, abutments and piers generally provide this supporting role whereas for long span suspension or cable stayed bridges, the cables and the towers function as the supports. The interface between a bridge deck and its substructures are the bearings, unless of course, the substructures are built into the deck. Finally, the most crucial items in the list of bridge supports are the foundations.

NEED FOR STRENGTHENING/MODIFICATION

3.　　Need for strengthening or modifying bridge supports may arise from various reasons. A few of these are identified below:-

(a)　　Change of function or use;

(b)　　Increased strength requirement;

(c)　　Design inadequacies.

CHANGE OF FUNCTION OR USE

4.　　Two examples are described here to illustrate the principles behind modification of bridge substructures arising from a change in the function of the obstacle over which the bridge was originally constructed. In the first example the case of an overbridge in the motorway widening scenario is considered. Currently in the United Kingdom many motorways and trunk roads are being investigated with a view to relieve traffic congestion. In many cases the likely result is that bridges constructed fairly recently will be demolished and replaced by new bridges to provide for additional capacity on the motorways. Under the circumstances it may be prudent to assume further requirements in the motorway facilities in the future.

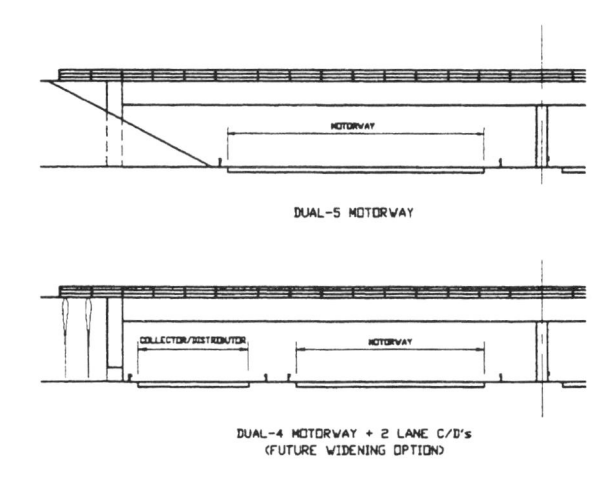

Fig. 1 - Modification of Abutments

5.　　In the second example the abutment is converted into a pier and an additional span is placed behind each of the original abutments to accommodate new facilities under the bridge. This situation is illustrated by photograph no. 1 which shows the elevation of the Bickenhill Lane Bridge north of Birmingham International Station.

PART 2: STRENGTHENING TECHNIQUES

Photograph 1 - Part Elevation Of Modified Bickenhill Lane Bridge

INCREASED STRENGTH REQUIREMENTS

6.   Increased strength requirements are generally caused either by changes in the Design Standards or by the introduction of new ones.

7.   Over the years there has been significant increases in the magnitude of loads used for the design of bridges. One only needs to compare HA (Highway) loadings as specified in BS153 Part 3A and BD37/88. The uniformly distributed load has increased by a factor of 2.02 for a 12m span and by a factor of 3.14 for a 900m span (loaded length) bridge. The last figure indicates one of the reasons why various strengthening works were needed in the Severn Bridge structure. The towers of this bridge were strengthened to substantially increase their load carrying capacity.

8.   The Building Research Establishment carried out a survey, in 1979, of the attitudes of bridge designers in the United Kingdom towards various aspects of the design of bridge foundations and substructures. The result of the survey is presented in a report[1]. The chapter on ABUTMENT EARTH PRESSURES AND STABILITY commences with the statement "The majority of designers use the concept of equivalent fluid pressure or nominal earth pressure coefficient when calculating earth pressures on the abutments of small and medium size bridges." The equivalent fluid pressure is of comparable magnitude to the "active" pressure as defined in CP2[2]. In fact, earth pressure which is much higher than the active pressure is relevant for the design of abutments and

retaining walls. Guidance given in CP2 was not adequate to cover all cases. Introduction of the Departmental Standard BD 30/87 has now clarified this particular aspect. The current bridge assessment programme in the United Kingdom does not require abutments of bridges to be quantitively assessed unless there is any sign of distress. In fact, significant underprovisions have been discovered as a result of assessment of some existing abutments.

9. Bridge piers located in highway medians and verges are increasingly being subjected to vehicle collisions. Photograph 2 shows the result of one such incident. An extract (Fig. 2) from a draft Departmental Standard, The Design of Highway Bridges and other Structures for Vehicle Collision Loads, shows the result of a Heavy Goods Vehicle Collision Study. The formal requirement to design highway bridge piers for collision loads was introduced first in 1973 through Department of Transport Technical Memorandum Bridges BE 5/73, Standard Highway Loadings and then in 1978 through BS5400 Part 2. However, the tonnage of heavy goods vehicles in the recent years has increased dramatically. As a result of this it has been necessary to reappraise the collision loads to be used for design and assessment of substructures. A comparison of values of collision loads introduced in

HGV COLLISIONS (1986 - 1990)

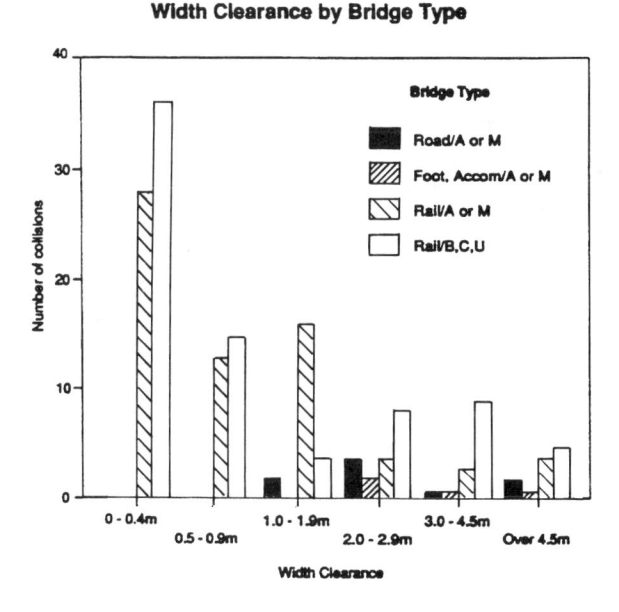

Width Clearance by Bridge Type

Fig. 2 - Distribution Of Bridge Support Collisions (England And Wales 1986-1990)

PART 2: STRENGTHENING TECHNIQUES

BS5400 Part 2 and those required by BD 48/93, The Assessment and Strengthening of Highway Bridge Supports, is given in Table 1. Manyfold increases highlight the reasons why it is not unusual to expect existing bridge piers to require strengthening to sustain the new requirements.

Photograph 2 - Collision Damage To Reinforced Concrete Pier

DESIGN INADEQUACIES

10. Inadequate provision for proper functioning of a bridge substructure may or may not immediately be apparent. Lack of detailed consideration of various aspects of foundation design may result in foundation failure which may take a considerable period of time to materialise. A regular inspection programme is essential to detect problems in time so that remedial measures can be taken.

11. Piers of river crossings may suffer from the effects of vessel collisions. Provision against collision damage in the original design may not be adequate. Any resulting failure can be quite dramatic generally resulting in the loss of many lives. Among many examples, the collapse of a major section of the Sunshine Skyway Bridge across Florida's Tampa Bay in 1980 caused by ramming of one of the piers of the bridge by the freighter Summit Venture (40,000 dwt), was perhaps the most spectacular. In 1983, the collision between a passenger ship and a railway bridge over the River Volga in the Soviet Union, resulted in the death of 176 people. Most recently in September 1993, 47 lives were lost in the railbridge disaster at Mobile, Alabama. All three disasters mentioned above were the direct result of vessel collisions with bridge piers. The necessity of impact protection of piers in navigable rivers cannot therefore be over-emphasised.

Table 1 - Comparison of Nominal Collision Loads on Supports Of Bridges Over Highways

BS 5400 : PART 2 : 1978			
	LOAD NORMAL TO THE CARRIAGEWAY BELOW, KN	LOAD PARALLEL TO THE CARRIAGEWAY BELOW, KN	POINT OF APPLICATION ON BRIDGE SUPPORT
Load transmitted from guard rail	150	50	Any one bracket attachment point or, for free standing fences, any one point 0.75m above carriageway level
Residual load above guard rail	100	100	At the most severe point between 1m and 3m above carriageway level
BD 48/93 : THE ASSESSMENT AND STRENGTHENING OF HIGHWAY BRIDGE SUPPORTS			
Main Load Component	500	1000	At the most severe point between 0.75m and 1.5m above carriageway level
Residual Load Component	250 (100)	500 (100)	At the most severe point between 1m and 3m above carriageway level

PART 2: STRENGTHENING TECHNIQUES

12. Bridges crossing watercourses are vulnerable also to damage resulting from scouring of the bed and thus undermining of the foundations. The problem has been highlighted in recent years by the collapse in the United Kingdom of two railway bridges. The first was the collapse in 1987 of a bridge at Glanrhyd in South Wales resulting in a train falling into the river. Four lives were lost. The second was in February 1989 when the 130 year old viaduct over the River Ness in the North of Scotland collapsed. Subsequent studies suggest that both failures involved scour of the bed material supporting piers of the bridges. There have also been reports in recent years of failures of road bridges in the United States resulting from undermining of pier foundations due to scour of the river beds.

STRENGTHENING/MODIFICATION TECHNIQUES

13. Over the years various techniques have been employed to modify and strengthen bridge supports and surely new techniques will be developed in the future to meet the needs of particular problems. In the remainder of the paper a number of examples will be used to illustrate techniques for modification of bridge substructures.

14. As shown in Fig. 1 the overbridge may, in the widening scenario, be constructed with buried abutments and embankment slopes in front of the abutment. This makes additional space available under the bridge for any future widening requirements by either removing the embankment fill in front of and attaching new wingwalls to the buried abutment or alternatively the soil nailing technique may be employed to retain the fill.

15. Modification of the Bickenhill Lane bridge is shown in photograph no. 1. The bridge was originally a single span bridge over the West Coast Main Line Railway. Construction of the Birmingham International Station necessitated additional lines under the Bickenhill Lane Bridge. This was provided for by converting the abutments into piers and extending the bridge with additional spans. The abutment wall as constructed did not have adequate space for the bearings of the side span beams. The problem was overcome by constructing a transverse beam, supported on the return wingwalls, which acted as a bearing shelf for the side span beams.

16. Strengthening or modification of an abutment with inadequate provisions to resist earth pressure and other forces, requires elaborate considerations. Fig. 3 illustrates a single span bridge with abutments having inadequate strength and a few of the options which may be adopted to bring the bridge up to design requirements. In (a), a practical method of strengthening the abutments is shown whereas in (b), (c) and (d), methods of reducing or fully eliminating earth pressure from the abutments are highlighted. The adoption of the Ground Anchors would in effect change the structural behaviour of the abutment and therefore knowledge of the structural details is essential. This is not the case with the earth pressure relieving methods.

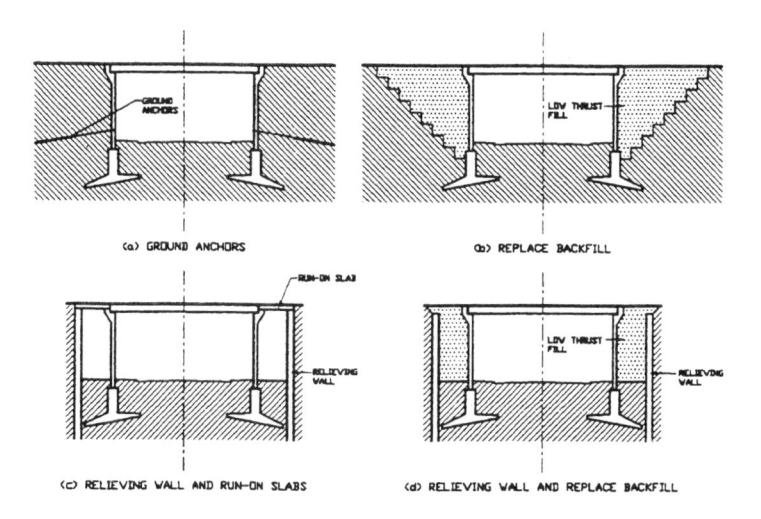

Fig. 3 - Modification of Abutments to Relieve Earth Pressure

17. Fig. 4 shows the bases of a bridge pier. An improvement of the junction required construction of an underpass through one of the spans of the bridge. The proposed underpass walls were very close to the pier bases. One of the bases was on strip foundation while the other was supported on piles. Both these bases required modification. To prevent any movement of the bridge pier as a result of the construction of the underpass, the loads from the bridge were transferred onto new piles. For the strip foundation, piles were constructed on two sides of the base, capping beams were constructed on the piles and the capping beam was then made integral with the base. Holes were drilled through the existing base and Macalloy prestressing bars were used to stress the pile caps onto the existing base. The base on piles needed modification because the existing piles were not long enough. The method adopted here was to construct new piles of length as required by the design and then cast a new base on the new piles but underneath the existing base so that the bridge load is transferred onto the newly constructed base. It was essential to cast a number of piles under the existing pile cap with a very limited headroom. The pier bases on the opposite side of the underpass were also modified.

PART 2: STRENGTHENING TECHNIQUES

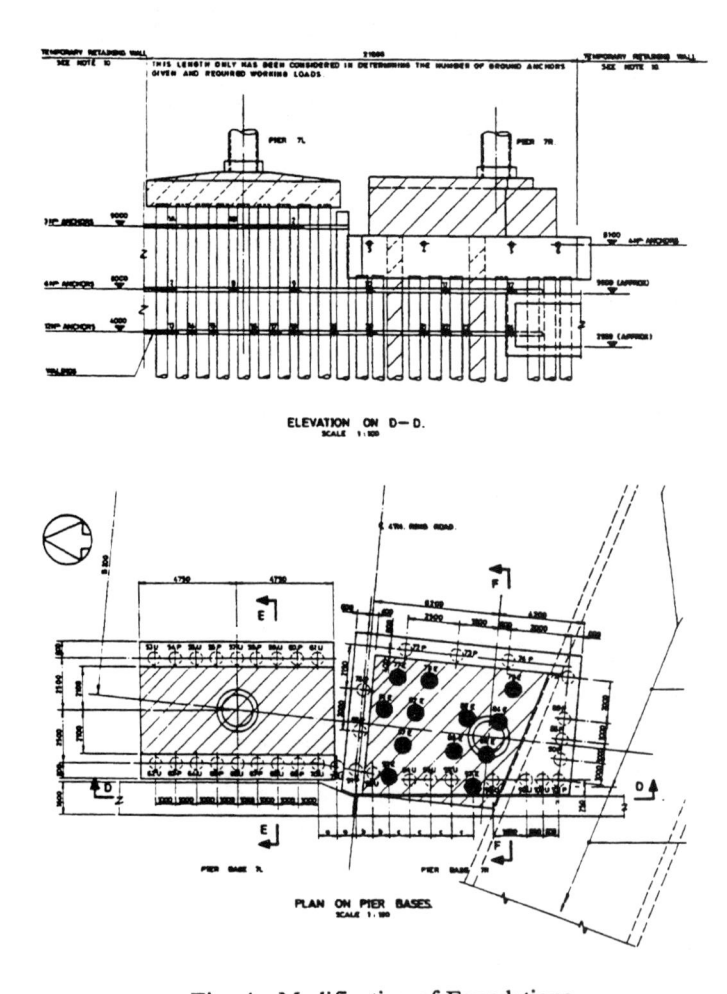

Fig. 4 - Modification of Foundations

18. As mentioned earlier in the paper, because of the recent introduction of large collision loads for the design of highway bridge piers, many existing piers will require strengthening in one form or another if the structures are to be brought up to current design standards. However, sometimes the design of an existing pier may be such that strengthening may be an impractical proposition. In this case it may be prudent to simply erect a concrete barrier to protect the pier from vehicle collision. The most recent application of this technique is the "Rapid Widening" Project, M4 Junction 4b to 5. Overbridge piers which were considered unsuitable for strengthening, have been protected by newly constructed concrete walls.

Photograph 3 - Strengthening of Piers

19. One form of strengthening a bridge pier comprising columns is to construct a concrete jacket around the columns. The jacket need not be carried right to the top of the column, its height being determined by the available strength of the existing column. The principle is illustrated in photograph 3 which shows recent strengthening of the columns of the Pease Pottage Interchange South bridge over the M23. Piers of three bridges were strengthened, in a scheme to widen the A23 from Pease Pottage to Handscross, using similar techniques.

20. Fig. 5 and photograph 4 illustrate an unusual case of bridge pier strengthening. The bridge involved is the Horsham Road Bridge over the A23 and the project is the A23 widening scheme as mentioned above. The columns of the pier are hinged at the junctions of their bases. Strengthening was required to resist vehicle collision loads and to support

PART 2: STRENGTHENING TECHNIQUES

an increase in dead load resulting from strengthening of the deck. The design of the bridge is such that it was essential to maintain the hinged connections. The base was widened with mass concrete to support a new prestressed concrete widened base and pedestal structure. Reinforced concrete collars were constructed around the columns to safeguard the columns against collision loads. A physical gap of 35mm around the existing column was formed by the use of polythene.

Photograph 4 - Strengthening of Bridge Pier Including Foundation

21. Fig. 6 illustrates a bridge pier which was assessed, under collision loads, as inadequate against shear failure at pile/cap interface and overloading of piles. The columns of the pier also require strengthening. In this case the columns will be strengthened by simply combining them to form a wall. The pile cap is proposed to be extended with extra piles to provide additional shear capacity and stability. The pier is in the verge and therefore minipiles, acting as tension piles, will be installed on the carriageway side. To provide the required shear resistances, two 900mm dia piles will be installed at each end of the base, beyond the bridge deck, because of the clearance requirements of large diameter piling equipment.

22. Strengthening of bridge piers against shipping impact is not a practical proposition. It is much easier for new designs to allow for navigation. Existing bridge piers can perhaps be protected by construction of barriers such as dolphins or in the extreme case artificial islands and by incorporation of electronic navigational aids.

23. In July 1990 the river pier of the Inn Bridge in Kufstein, Austria, suffered damages from foundation settlement. In less than 24 hours the downstream side of the pier sunk by 1.15m while the upstream side sunk

by 0.22m. In addition, the pier was also leaning towards the right bank. There were damages to the bridge deck and other substructures. The displacements of the river pier were triggered by erosion of the river bed, the deepest scour being located at 4.5m below the base of the pier foundation. Measures to rehabilitate the pier included placement of some 40,000t of rock to stabilise the pier initially and finally a combination of high pressure grouting and construction of 136 jetting piles 1.2m in diameter. Further details can be obtained from reference (6).

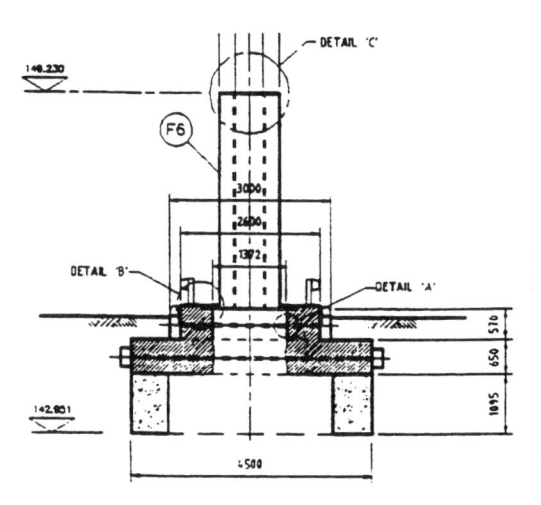

Fig. 5 - Strengthening of Bridge Pier Including Foundation

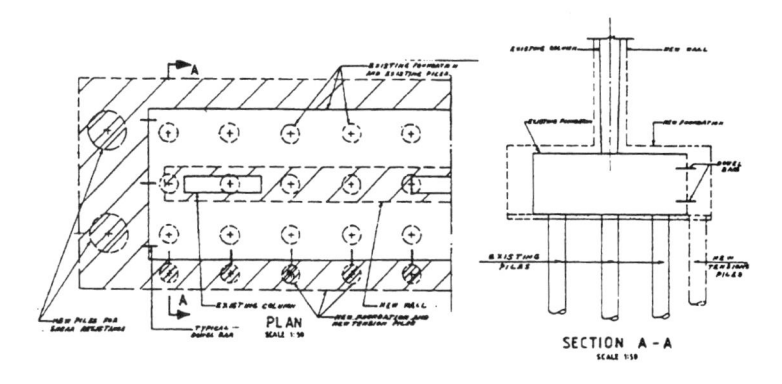

Fig. 6 - Strengthening of Bridge Pier Incorporating New Piles

24. Sound provision against scour damage of bridge substructures have not always been made in the past. Bridge failures at Glanrhyd and Inverness led British Rail to initiate a scour risk evaluation programme, engaging University of East Anglia to investigate methods of reducing the amount of scour at bridge piers. The object was to provide a cheap but effective way of improving the safety of the foundations. The results of

the research was published in a paper by Paice et al[9]. The solution arrived at was to install a group of small width piles, relative to the width of the pier, in effect to relocate the scour location away from the bridge pier. Early in 1993 the method was being tested out at three separate river viaducts. Successful protection of bridge supports have also been achieved in less severe inland waterway situations by placement of stones of suitable size and weight, in certain cases, underlain by bedding layers and geotextile membranes.

ACKNOWLEDGEMENTS

25. The author sincerely thanks all his colleagues in Acer and officials of the Department of Transport who have assisted in the preparation of this paper. Some of the work described in the paper originates from Acer's work with the Department of Transport, South Eastern Construction Programme Division and the author is thankful to the Director (SECPD) for his permission to include such material in this paper.

REFERENCES

1. Building Research Establishment - Bridge Foundations and Substructures, HMSO 1979.

2. Civil Engineering Code of Practice Joint Committee - CP2 : Earth Retaining Structures, 1951.

3. Department of Transport - Backfilled Retaining Walls and Bridge Abutments, Departmental Standard BD 30/87.

4. Department of Transport - The Design of Highway Bridges and Other Structures for Vehicle Collision Loads.

5. Department of Transport - The Assessment and Strengthening of Highway Bridge Supports, Departmental Standard BD 48/93.

6. W Kittinger and M Ashabar - Inn Bridge, Kufstein, Austria : Bridge Damage and Repair. BRIDGE MANAGEMENT 2 Thomas Telford, April 1993.

7. New Civil Engineer - RETROFIT PROTECTION CALL AFTER US CRASH, 30 September 1993.

8. J Riddell - Problems Associated with Assessing Existing Bridge Structures for Scour Failure. BRIDGE MANAGEMENT 2, Thomas Telford, April 1993.

9. C Paice, R D Hey and J Whitbread - Protection of Bridge Piers from Scour. BRIDGE MANAGEMENT 2, Thomas Telford, April 1993.

Crossbeam replacement

M. S. CHUBB, Principal Engineer, and I. L. KENNEDY REID, Senior Group Engineer, W. S. Atkins Consultants Ltd

SYNOPSIS.
The deterioration of bridge substructures caused by corrosion induced by deicing salts, is a common problem in the UK's bridge stock. Generally piecemeal repairs are carried out but complete replacement is preferable in terms of the durability of the final product. This paper describes the development of a scheme for replacing a motorway crossbeam support from concept to successful completion. The problems encountered especially with respect to undertaking the work with minimum traffic disruption, are described in detail. Similar projects are reviewed and recommendations made for further developments.

THE PROBLEM
1. The Midland Links Motorway Viaducts carry the M5 and M6 Motorways around the suburbs of Birmingham and comprise over 1200 spans of elevated structures. These are generally simply supported steel and concrete composite bridge decks supported by reinforced concrete crossbeams and columns.
2. De-icing salts have leaked onto the crossbeam supports through leaking joints, causing such widespread corrosion that the majority have to be repaired. A major maintenance programme is underway with most of the crossbeams being repaired and cathodically protected to prevent further deterioration.
3. However, some crossbeams have been found to be in such serious condition that they were considered to be almost unrepairable. A scheme was therefore developed to replace crossbeams completely and to carry out this operation with minimum disruption to motorway traffic.

THE OBJECTIVE
4. The crossbeam which required replacement carried the dual three lane M5 Motorway which has typical traffic flow in excess of 65,000 vehicles per day. The crossbeam was 33 m long, 1.68 m wide and 1.52 m deep supported by two 1.52 m diameter columns at 18.3 m centres. One column 8.3 m in height is supported by a 6.0 m diameter spread footing founded on stiff clay. The other column 5.1 m high is founded on a 6.0 m diameter base supported by 16 No. piles end bearing on dense gravel or stiff clay. The decks are 15 m span, simply supported, each with 10 No. steel universal beams acting compositely with in-situ reinforced concrete slabs. Reinforced concrete panel walls connect the crossbeam directly to the deck slab providing transverse and longitudinal restraint to the superstructure and precluding the need for bearing stiffeners. The bearings were sliding steel on steel with a steel rocker providing longitudinal rotational capacity. The basic configuration of the deck and supports is indicated in Figure 1.

Bridge modification. Thomas Telford, London, 1995

PART 2: STRENGTHENING TECHNIQUES

5. The crossbeam was considerably deteriorated due to corrosion of the top and bottom reinforcement caused by a combination of chloride contamination and lack of cover on the soffit. There was approximately 50% delamination of the concrete surface and considerable loss of section of links and main bars. If the beam had been repaired by conventional methods of replacement of contaminated concrete, then 90% of the surface area would have had to be removed. For cathodic protection then all the delamination would have had to be repaired. The main reason for replacement, however, was the condition of the reinforcement and the practicability of repair. The basic requirement was therefore to remove the crossbeam completely and to construct a similar replacement. This work would have to be carried out with the minimum of traffic disruption.

6. In the case of this particular crossbeam it was located adjacent to a canal on one side and an embankment on the other side both of which interfered with access and affected the structural solutions. See Figure 1.

SAFETY AND RELIABILITY

7. An over-riding consideration in the design and execution of this scheme was the safety of the deck structure because of the consequences of affecting traffic on the M5 Motorway. When the permanent supports were removed then the temporarily supported structure was much more vulnerable to accidental problems. Whilst catastrophic collapse was the major consideration, with possible loss of life, the economic costs would have been enormous. If the M5 Motorway had to be closed, then the resulting traffic delay costs would be of the order of £1M per weekday. For this reason the scheme had to take account of unforeseen difficulties. Consequently a highly redundant design was preferred in order to increase the reliability of the structure.

THE OPTIONS

8. A number of schemes were investigated at the feasibility stage. These are described below and are shown in Figure 2.

Deck Supports

9. The most obvious scheme, involved supporting the decks each side of the crossbeam leaving access for demolition and reconstruction. The temporary supports however required independent foundations.

Beam to One Side

10. This scheme involved constructing a new crossbeam and foundation on one side of the existing crossbeam with a halving joint to support the far span. Although there would have been considerable technical difficulties with the halving joint and strengthening the shear connectors on the existing beam, there were attractions as there was less risk. Temporary supports were not required and the new support would not be contaminated by leaking joints in the future.

11. A variation of this scheme involved constructing temporary supports on one side so giving good access to demolish and reconstruct the crossbeam.

Beam Under

12. This was a modification scheme rather than replacement and whilst it was attractive in terms of cost it did not achieve the objective of eliminating the present deteriorated crossbeam.

Conclusion

13. The principle of the first scheme was considered the most suitable and was therefore taken forward to the detailed design stage and is described in more detail in the following sections.

TEMPORARY STEELWORK SUPPORTS

14. The temporary supports were required to cantilever over the canal on one side and oversail an embankment slope on the other. The scheme devised consisted of a steel frame with inclined legs cantilevering over the canal with a large concrete counterweight. See Figure 3.

15. Transverse stability of the main beam members was achieved by pairing up the frames and interlinking them with tie beams which provided a suitable seating for both the jacks and the temporary pot bearings. This lent itself to support by twin columns. See Figures 4 and 5. The frames were also required to deflect with longitudinal temperature movements of the deck as they relied on longitudinal fixity for their stability in that direction.

16. Deck beam restraint brackets were fastened between the temporary supports and the deck beams to provide longitudinal and transverse fixity during jacking. After jacking the temporary pot bearings provided this fixity.

17. Additional safety was provided by shimming the crossbeam directly under the deck beams enabling the deck and limited live loading to be carried by a single leaf of the frame in the event of damage to the columns of the other leaf. In such an event the deck beam restraint brackets would again provide longitudinal and transverse fixity.

18. The temporary supports were founded on reinforced concrete slabs bridging between additional piles bored either side of the existing foundations. The bridging slabs were designed to carry possible propping of the existing beam during demolition and falsework for construction of the replacement beam.

MODIFICATIONS TO THE EXISTING STRUCTURE

19. In order to jack the structure off its existing crossbeam support it was necessary to modify the structure for a number of reasons.

Bracings

20. Firstly, the panel walls connecting the crossbeam directly to the deck slab had to be removed otherwise the deck could not be separated from the crossbeam. As the panel walls provided transverse stability and longitudinal restraint as well as support to the ends of the decks, they were replaced by K-bracing as is shown in Figure 6. Positive support to the ends of the slab deck was provided by pumping grout into special bags placed between the top bracing member and the deck soffit as is shown in Figure 7.

Bearings

21. An additional set of temporary bracing was also required to stabilise the deck at the points of temporary support. These are shown in Figure 8.

22. In the final structure the existing steel on steel bearings were replaced with conventional pot bearings in order to provide transverse rotational capacity at the new bearing stiffener locations.

Shock Transmission Units

23. The 'floating' articulation was changed to fixed/free, and in order to share longitudinal loads between bents, shock transmission units (STU's) were provided. These were also necessary for the temporary propped condition in order to share longitudinal loads between the adjacent bents. See further discussion on viaduct articulation below.

24. The STU characteristics were specified following calculations of the time for which traction loads would be applied. The movement of the STU had to be sufficiently small during the period of application of the traction load to sustain sufficient resistance between the deck ends. The resistance of the STU to temperature movements, given their extended period of application, had to be sufficiently small not to overstress the viaduct supports.

PART 2: STRENGTHENING TECHNIQUES

25. The STU's were pinned to brackets welded to large plates bolted to the underside of the deck. Adequate tolerance had to be permitted in the bolt location to avoid the reinforcement in the deck. The plates oversailed the trimmer beam of the K-bracing and transmitted shear forces and moment to the deck. They had to be braced to provide sufficient strength. See Figure 10.

TEMPORARY GUIDES TO BEARINGS

26. The removal of the panel walls released the transverse restraint to the deck. Until the deck had been lowered on to the transversely fixed bearing on the temporary supports, there would have been no transverse restraint to the deck. This problem was overcome by fixing temporary guides to selected existing bearings prior to demolishing the last panel walls. The guides had to operate both before and during the jacking operation until the bolts of the fixed temporary bearing were tightened.

27. Angles with machined faces were therefore bolted to the deck beams, and brackets, which were clamped to the existing bearings, fitted against them to provide guiding faces which would slide longitudinally and vertically while transmitting the required transverse loads. The sliding surfaces were greased, and the brackets were released in a controlled manner at the end of the jacking operation to transfer the transverse forces from the existing bearings to the temporary fixed bearing.

DYNAMIC BEHAVIOUR

28. During the design there was concern that there may be dynamic problems with the behaviour of the deck when temporarily supported. With traffic crossing the structure it was possible that 'springboard' action of the deck cantilevers may cause excessive vibrations.

29. Dynamic analyses were carried out using a special computer program which allowed the effects of a vehicle crossing the deck to be modelled. Results were given in terms of deflections, forces and reactions plotted against time. A typical plot of temporary support reaction and cantilever deflection is shown in Figure 11 which showed that the deck could uplift at the temporary jacking point and considerable vibrations would occur at the cantilever tip.

30. In order to avoid the problem, a pin was introduced which linked the two deck ends together. The change to the dynamic behaviour is shown on the other plots in Figure 11. The pinned connection units (PCU's) were fixed to the deck beam ends by substantial brackets which were designed to avoid the new bearing stiffeners. Slotted holes were used in one of the brackets to cater for construction tolerances and deck beam deflections between installation prior to jacking and bolt tightening after jacking. The brackets had to be appropriately shimmed to allow for lack of alignment between the pairs of deck beam ends. Since the deck beam webs were not truly vertical, and to cater for small torsional rotations of the deck beams, spherical joints were chosen between the vertical link and the brackets (see Figure 9). These required periodical greasing while in use, and tubes were led away from the joints to more accessible greasing points. The PCU's were subjected to a full scale laboratory loading test prior to installation.

31. The other possible solution of changing the support points to reduce the cantilever lengths was rejected because of the reduced access space for work on the crossbeam and deck. This is discussed later in Section 54.

VIADUCT ARTICULATION

32. Although the viaducts comprise simply supported spans, their articulation is unusual and also highly indeterminate. The steel deck beams rest on steel on steel sliding bearings which were initially greased during construction but are known to now have high friction coefficients approaching unity. The decks are stabilised by the reinforced concrete panel walls which connect the deck slab directly to the crossbeams. Transverse forces are directly transmitted from the deck to the crossbeam while in the longitudinal direction the

panel walls flex to accommodate temperature movements but are capable of transmitting any longitudinal forces which are in excess of the bearing friction forces.

33. A single span in a long length of similar spans acts in a similar manner to continuous welded rail, as the adjacent spans offer restraint and temperature movements are taken up in the span module by sliding, even on the high friction bearings. When the articulation is altered by introducing free sliding bearings then the modular system no longer applies and interaction occurs. If the continuous welded rail analogy is considered then the effect is similar to cutting the rail as large movements occur where previously there were none.

34. Whilst some movement was expected, the amount was considered unpredictable although attempts to model the effect using non linear analyses were later carried out with some success. When the structure was monitored on site the expansion and contraction lengths were found to be equivalent to the thermal movement of four to five span lengths which under extreme temperature conditions could have overstressed the temporary supports. In order to control the movements to within acceptable limits, restraints were added to the deck beam ends to give the continuity which had been removed by the articulation modifications. These consisted of tie bars to transmit tensile forces during contraction and packs between the bottom flanges to transmit compression forces during expansion.

ERECTION AND INSTALLATION OF SUPPORTS

35. The deck jacking was to be carried out with the Motorway closed to traffic over a weekend. The restricted period available for closure of the Motorway required the jacking procedure to be carried out as efficiently as possible. In order not to overstress the deck slab or the K-Bracing, the deflections of the deck beams had to be kept to within +/- 1 mm of the adjacent beams. Thus at each stage of jacking the deflections of each of the 20 deck beams had to be read. It would have taken too long to read each of 20 dial gauges positioned at each deck beam, therefore, the jacking procedure specified the use of linear variable displacement transducers (LVDT's) at each beam, led back to a central console adjacent to where the 40 jacks could be pressurised. In the event, this procedure worked very well.

36. In jacking up the deck, a nominal gap had to be achieved over the existing bearings to prevent live load closing the gap, and to permit the existing bearings to be removed. As the deck was changing from being simply supported on the existing bearings to cantilevering beyond the jacking points, the required lift at the jacks was greater than the required gap over the existing bearings. This was calculated, but as it involved the concrete deck, acting compositely with the deck beams, going into tension, the accuracy of the calculation could not be guaranteed. By jacking up the deck at a rehearsal prior to the final operation, the lift at the jacks, and hence the thickness of the spacer plates required over the temporary bearings, could be determined. This then saved time during the final jacking, by streamlining the operation of inserting the spacer plates. The other equally important reasons for the rehearsal were to practise and time each operation to ensure it could be done in the time available, and to ensure that the deck could be lifted satisfactorily on the jacks without tilting from side to side, or losing synchronisation between the lifting of the two deck ends. In the event, the jacking rehearsal was to prove worthwhile in ensuring that for the final operation the motorway was reopened in time, prior to the heavy Monday morning traffic.

37. The levels of the temporary steel supports were carefully monitored as the deck load was applied to ensure that the structural behaviour was as anticipated.

38. The procedure for the jacking had to be laid down carefully for the Contractor to follow, to ensure that each item described above was completed at the appropriate point during the jacking operations. There were many meetings with the Contractor and specialist jacking sub-contractor to ensure that the operation could be carried out within the possession periods. This led to some changes which streamlined the operations.

PART 2: STRENGTHENING TECHNIQUES

SAND LORRY TRIALS

39. In order to ensure that the PCU's operated satisfactorily in eliminating unacceptable vibrations in the deck beams, immediately after jacking a sand lorry was run a number of times across the deck joint at differing speeds. A timber plank was positioned on the carriageway for several of the runs to simulate an impact loading. The amplitude and frequency of the deck deflections were recorded by accelerometers, at mid span, cantilever end, and over the temporary supports.

40. Vibration monitoring was continued under live traffic to provide a further guarantee. The results of the monitoring were compared with the theoretical dynamic analysis of the deck before proceeding with the demolition of the crossbeam. In the event the theoretical results matched well with those in practice and the effectiveness of the PCU's was confirmed.

DEMOLITION AND RECONSTRUCTION

41. The demolition involved the removal of 60 m^3 of concrete and 30 tonnes of reinforcement. There were considerable constraints of access between the temporary supports and there were safety considerations as large pieces of reinforced concrete were to be removed and the Contractor wished to use machine mounted breakers.

42. It was decided not to remove the crosshead in pieces as this would present too many problems with handling and impose risks of damage to the temporary supports. The use of machine mounted breakers was tried by the Contractor but this was found to cause large vibrations in the columns and foundations which were to be retained.

43. The Contractor therefore used hand held breakers in conjunction with mechanical disruption devices which initially cracked the concrete.

44. Where reinforcement had to be retained, eg, the column starters, then water jetting was used. This was successful, though time consuming. The core concrete was found to be particularly difficult to remove because it was strong and heavily reinforced.

45. The Contractor laid his crossbeam soffit formwork in advance and used it to collect the demolition debris. On completion of the demolition this was adjusted to line and level and the reinforcement cage assembled in-situ. Because of the site constraints and the protrusion of the column starters prefabrication of reinforcement was not possible.

46. With the side forms in place access was again restricted and the high slump 10 mm aggregate concrete was placed by pump. A trial of the concreting operation was carried out to confirm the mix suitability and to ensure that compaction could be carried out successfully. This proved to be a worthwhile exercise.

47. The permanent pot bearings were set at levels which allowed for elastic and creep deformation of the replacement crossbeam. The outermost bearing were set high to minimise the creep deflection loading on the K-bracing.

DEJACKING

48. The dejacking was again carried out under a complete motorway possession. This was generally a reverse of the jacking procedure. There were a considerable number of activities to be carried out and the programme was carefully planned and executed. After dejacking, the lower connections of the K-bracing were released and rebolted to relieve load arising from any construction tolerances in the levels of the permanent bearings.

FURTHER APPLICATIONS

49. Crossbeam replacement was always regarded as a last resort but the exercise was useful in identifying problems and allowing the costs to be realistically assessed. Compared to the other options for crossbeam repair then there are clear disadvantages. However, providing a new uncontaminated structure with minimal future maintenance also has its attractions.

50. In reviewing the past work there are improvements and alterations which could be made to reduce the large costs and these are explained below.

Simplifications

51. The supports could be simplified by reducing the onerous safety requirements. Instead of a double leaf system then a single leaf could be used, as has been used previously on a similar application. The structure however is vulnerable if one of the legs is damaged in which case collapse would occur. Given the knowledge of site operations and the protection measures which can be introduced this may be viable but whether the increased risk is acceptable is debatable.

52. By incorporating bearings on jacks or only using jacks the lifting and dropping down operations could be avoided.

Foundations

53. Providing foundations for the temporary support was difficult, time-consuming and expensive. A scheme has been developed which uses the existing foundations for support but this would have proved very complex to construct and erect underneath the crossbeam.

Other Modifications

54. In order to overcome the dynamic problems with the cantilevers then the support points could be moved by providing beams spanning between the temporary supports and supporting the deck from cross girders which form part of the permanent bracing. This has advantages by eliminating the use of pins connecting the two deck ends but makes access to the crossbeam even more difficult. It has been successfully used elsewhere however.

Jacking Under Traffic

55. Further applications must avoid the traffic disruption of motorway closures during jacking operations and so live load jacking will be necessary. Given the knowledge gained from the previous work this is considered quite feasible and two schemes are being developed. One involves jacking the structure with the pins already connected and the other involves damping the free ends of the deck using jacks and rubber bearings.

CONCLUSIONS

56. The removal of a complete motorway support under live traffic conditions was successfully carried out although it proved to be a complex and expensive operation. The major considerations are the safety of the structure because of the massive economic costs of disrupting traffic on a major highway.

57. Some of the problems associated with the design and construction have been described and improvements have been considered for future applications.

ACKNOWLEDGEMENT

58. The paper is published with the permission of the Department of Transport West Midlands Network Maintenance Division. The authors acknowledge the assistance of all colleagues involved in this project including the independent checkers G Maunsell & Partners, the structural auditors Flint & Neill Partnership and the main contractor Tilbury Douglas Construction Limited.

PART 2: STRENGTHENING TECHNIQUES

Fig. 1. General Arrangement of Crossbeam

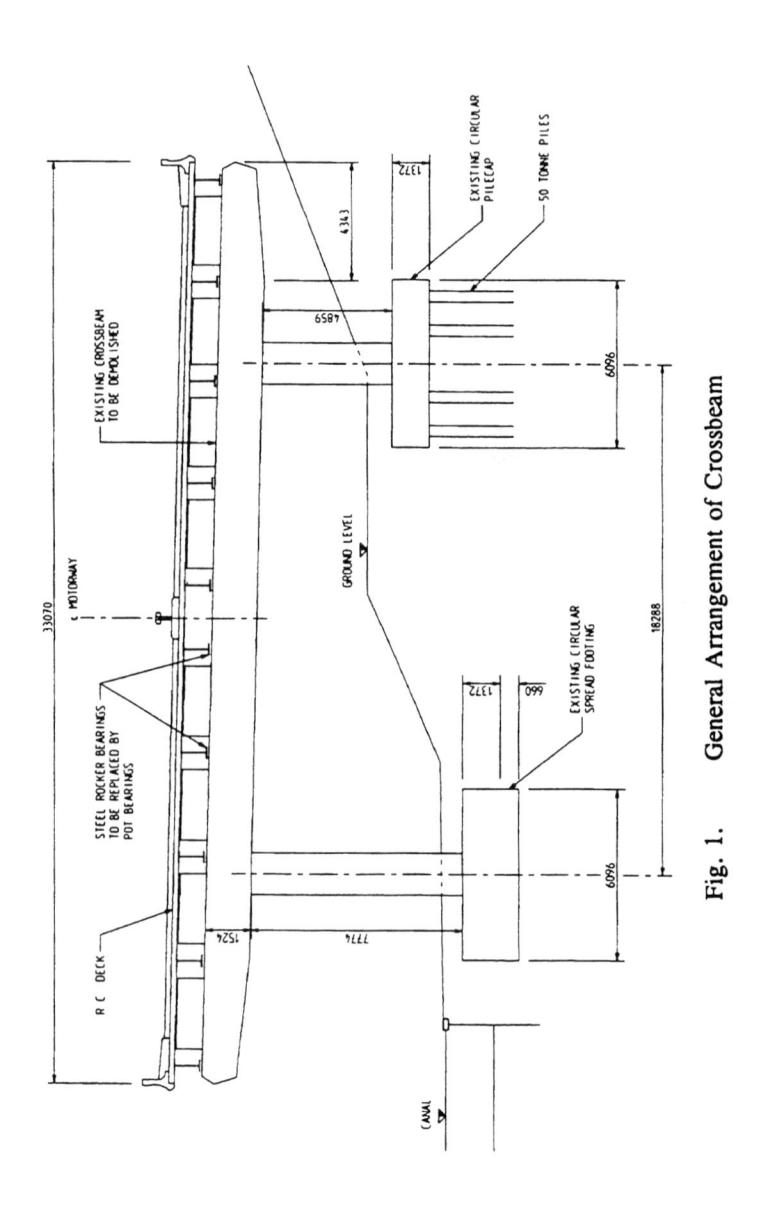

248

A. JACKING DECK ON BOTH SIDES

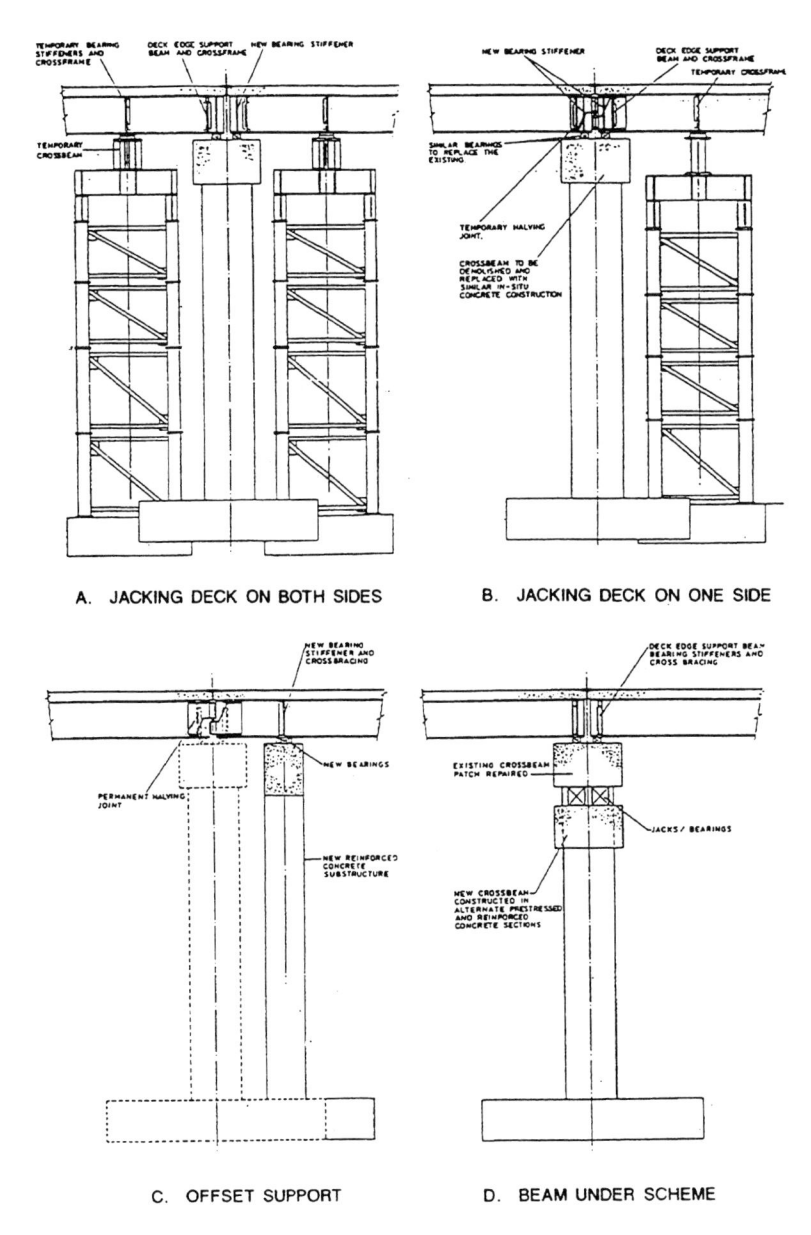

B. JACKING DECK ON ONE SIDE

C. OFFSET SUPPORT

D. BEAM UNDER SCHEME

Fig. 2. Alternative Schemes for Crossbeam Replacement

PART 2: STRENGTHENING TECHNIQUES

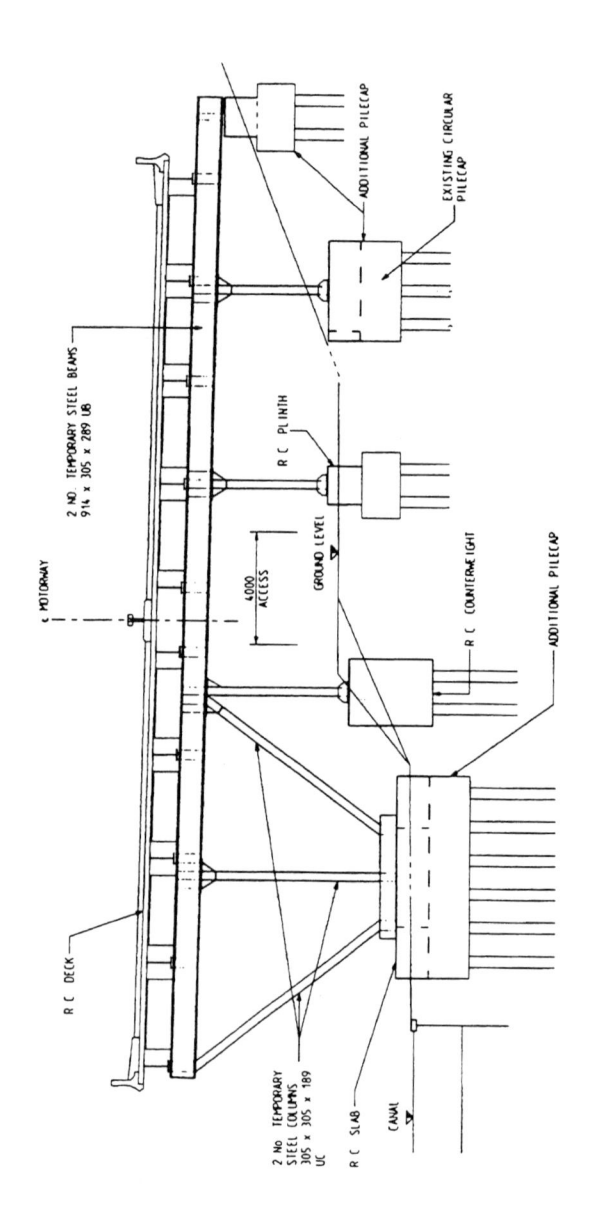

Fig. 3. General Arrangement of Temporary Support

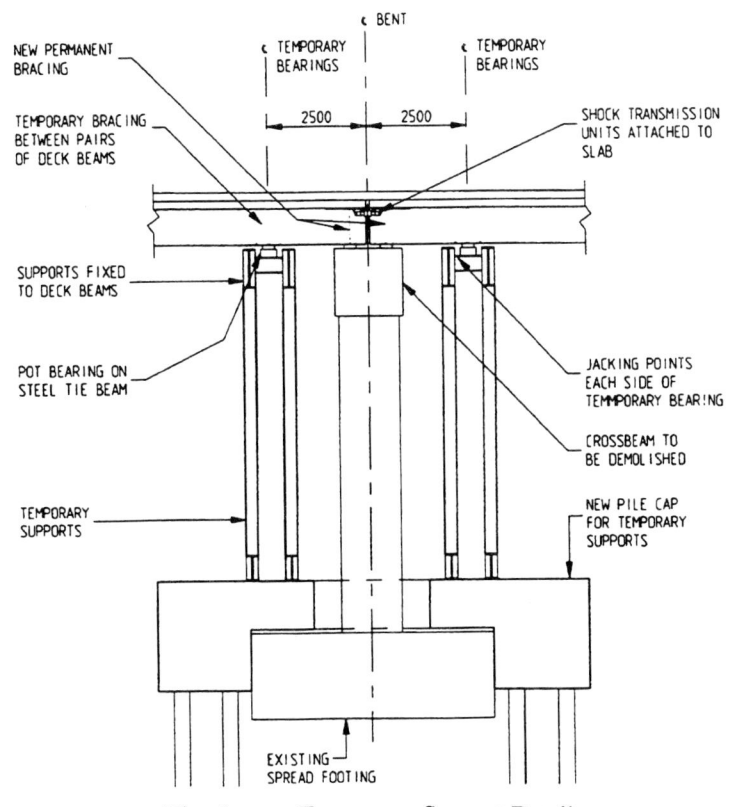

Fig. 4. Temporary Support Details

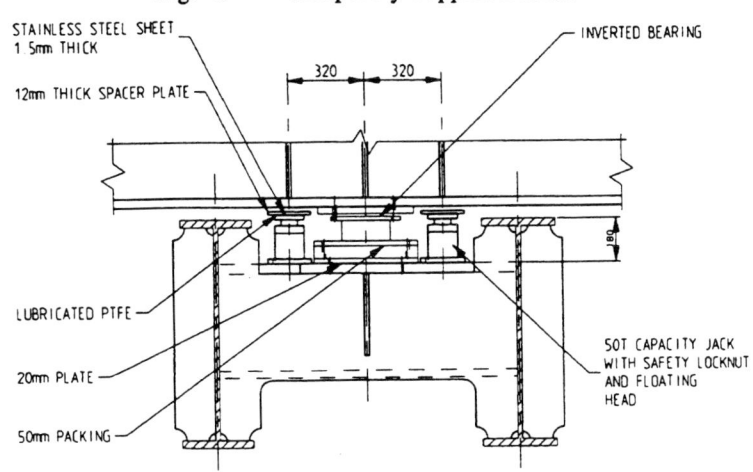

Fig. 5. Jacking and Temporary Beam Details

PART 2: STRENGTHENING TECHNIQUES

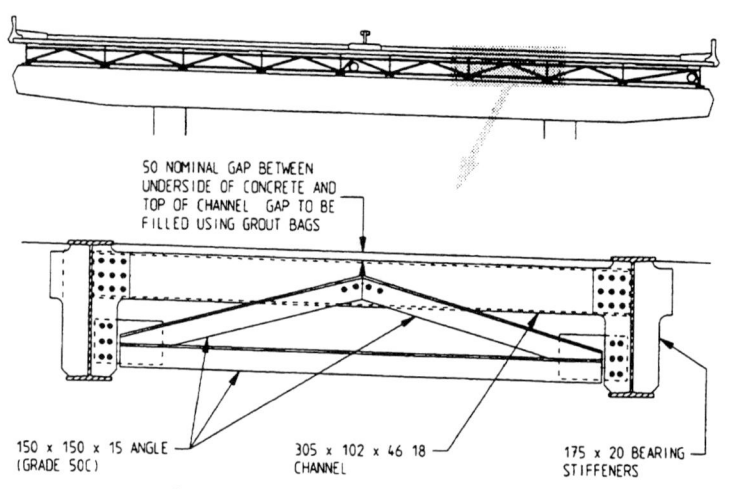

Fig. 6. Permanent Bracing Details

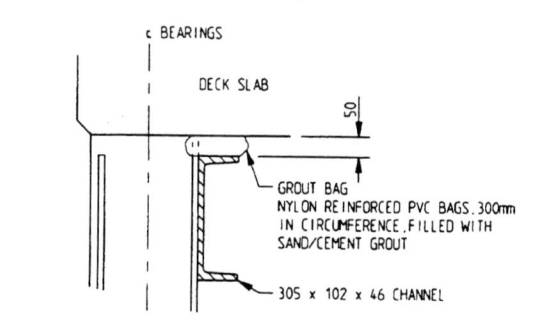

Fig. 7. Grout Bag Detail

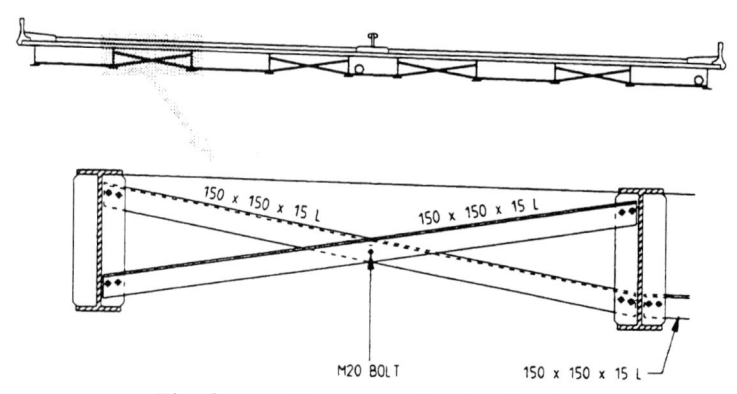

Fig. 8. Temporary Bracing Details

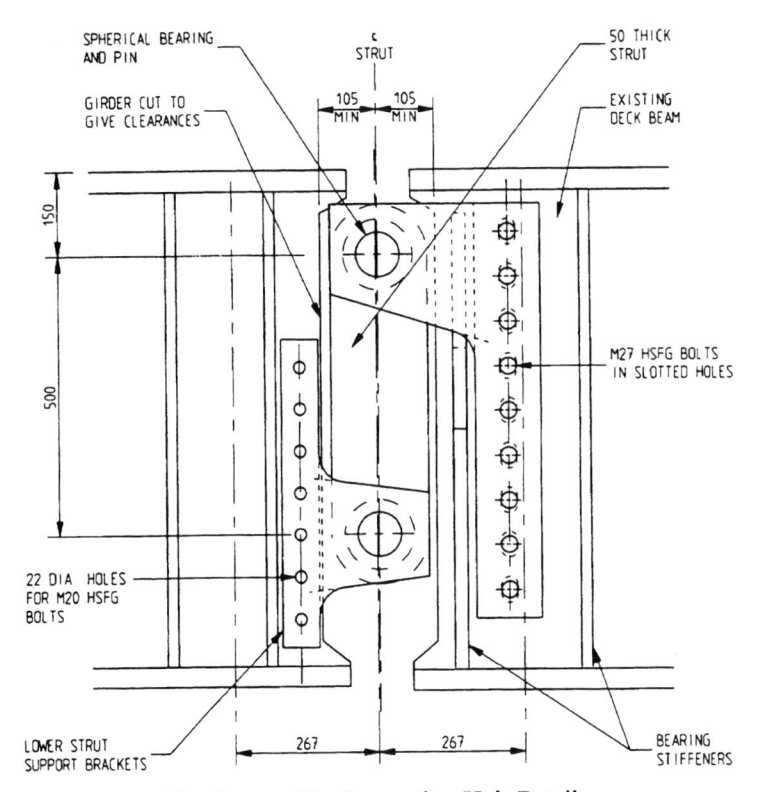

Fig. 9.   Pin Connection Unit Details

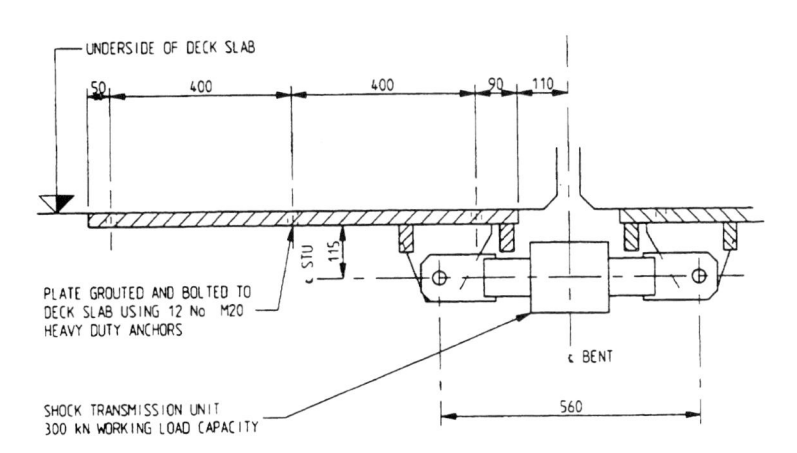

Fig. 10.   Shock Transmission Unit Details

PART 2: STRENGTHENING TECHNIQUES

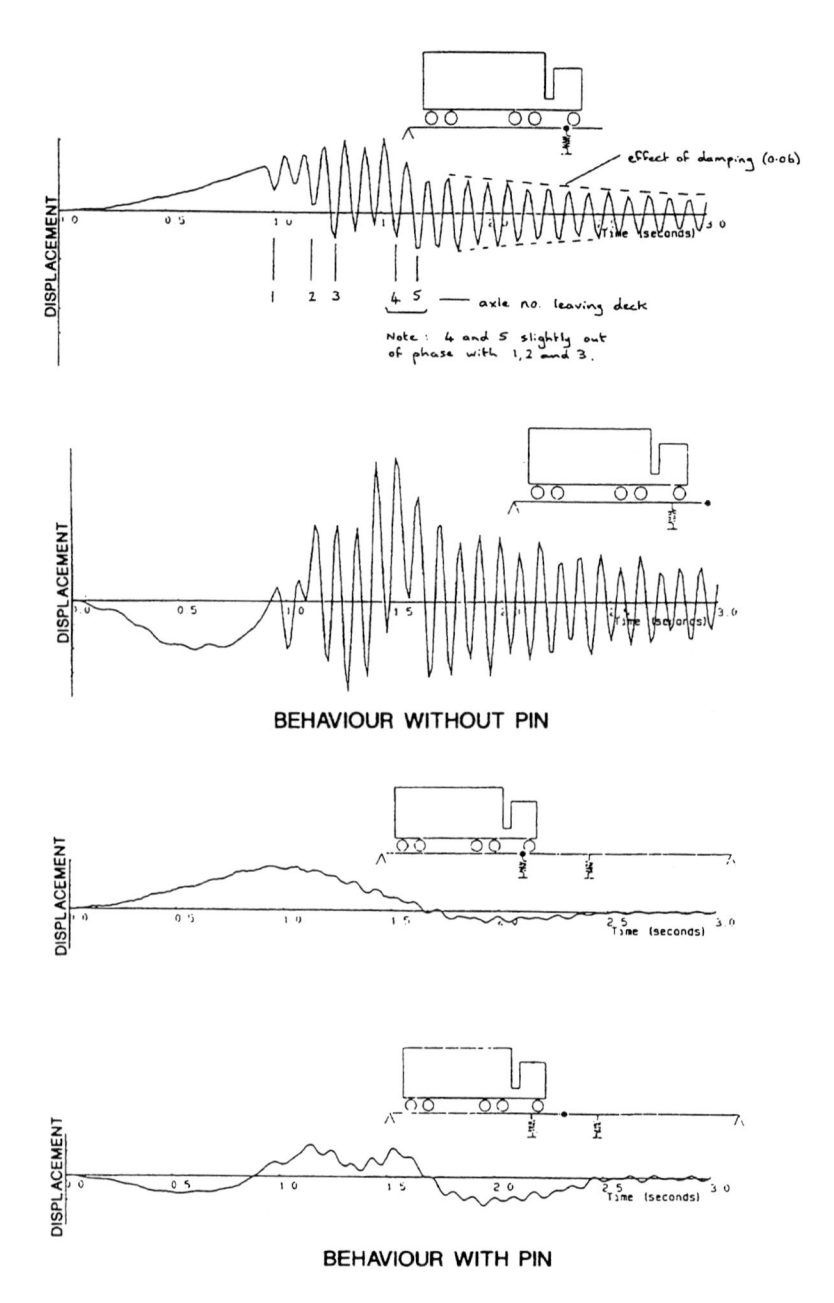

BEHAVIOUR WITHOUT PIN

BEHAVIOUR WITH PIN

Fig. 11.   Dynamic Analysis Results

Use of higher containment vehicle restraints (safety fences and barriers)

C. WILSON, Principal Engineer, Bridges Engineering Division, Department of Transport

SYNOPSIS. Over the past decade, higher containment vehicle restraint systems such as safety fences, safety barriers and bridge parapets, have been developed in both metal and concrete. This Paper explains this research work and gives basic details on how the deployment of such systems, particularly the permanent and temporary higher vertical concrete safety barriers, are proving to be a very effective and economical method of providing protection to structures and containment to withstand current collision load requirements.

BACKGROUND

1. Since their general introduction in the mid 1960's, vehicle restraint systems, such as safety fences and parapets, have normally been designed and tested to contain and safely redirect a 1.5t car (i.e. a medium to large saloon), travelling at 113 km/h and impacting the restraint at an angle of up to 20°. This class of vehicle represents over 70% of traffic on our motorways and major all-purpose dual carriageways.

2. Over the past 10 years, although the total traffic mileage of heavy goods vehicles has remained fairly constant, the number of 4 and 5 axle articulated goods vehicles has increased very significantly. In 1983, the authorised maximum gross weight for HGVs was increased from 32.5 to 38t and now HGVs with a gross vehicle weight over 25t account for nearly 30% of all HGVs.

3. It was at this time the Department of Transport started to examine the development of stronger safety fences, safety barriers and also bridge parapets. Since then the Department, through its research arm TRL, and with the involvement of Industry, has successfully developed and introduced restraint systems capable of containing a range of vehicle impacts from a Mini (0.8t) at 113km/h to a 39.2t articulated truck at 80km/h and a 30t rigid tanker at 64km/h all impacting at 20°.

4. This latter impact condition is the 'P6 level of containment', specified in Departmental Standard BD 52/93 (ref.1) for bridge parapets over busy railways and at other locations where the consequences of such a vehicle leaving the bridge could be extremely hazardous.

PART 2: ALTERNATIVES TO STRENGTHENING

5. Initially, the impact performance specification selected for higher containment safety fences and barriers was a 16t two-axled rigid lorry travelling at 80km/h but, more recently, consequent to the introduction of revised collision load requirements for structural supports, the 30t x 64km/h x 20° impact conditions have been adopted to match the requirements for bridge parapets.

HIGHER CONTAINMENT STEEL SAFETY FENCING.

6. <u>Single sided, double rail Open Box Beam (SS,DROBB).</u> From the initial R&D work undertaken in the 1960's, a safety fence, the single sided, double rail open box beam (SS,DROBB), which has a 5t x 80km/h containment level, was introduced. For the past 25 years, this type of fence (Fig.1) has been installed at many locations on the highway network where potentially hazardous conditions exist (e.g. adjacent to large areas of deep water or fuel storage tanks etc.). Details of the SS,DROBB are in the Highway Construction Details (ref.2).

Fig. 1. Single Sided, Double Rail Open Box Beam Safety Fence

7. <u>Double sided, double rail Open Box Beam (DS,DROBB).</u> Based on the general successful use of this system, the research for a higher containment safety fence used the same basic design concept but two additional box beam rails on the reverse side were introduced to form the double rail, double sided open box beam safety fence (DS,DROBB). This fence (Fig. 2) consists of four parallel box beams set in pairs each side of the fence at centre heights of 0.610m and 1.02m. The beams are supported on 125 x 90mm 'Z-section' knock down posts with spacers and extra cross bracing stiffeners at mid-span to hold the beams in position during impact.

Fig. 2. Double Sided, Double Rail Open Box Beam Safety Fence.

8. To overcome potential problems of violent longitudinal decelerations of the smaller car, a lower hollow steel section rubbing rail was added to reduce the potential for the wheels of an errant car to make contact with the 'Z' posts. Otherwise it was designed to use, as far as possible, the standard open box beam safety fence components.

9. This safety fence contained and redirected cars impacting at angles up to 25° at 113km/h and although a 30t 4-axle rigid HGV overturned the fence it successfully contained HGVs up to 38t travelling at 80km/h at an angle of 15°, and a 16t vehicle at 25° with a speed of 64km/h. Large lateral displacements of the fence beams occurred, however, when the impact involved a goods vehicle.

10. The higher containment safety fence (DS,DROBB) has so far been installed at one location in England and this is in the eastbound verge on the M6 motorway (West of Jct 3) in front of the Corley Service Area Restaurant building. This facility is located fairly near the motorway boundary and only slightly above carriageway level. A slender pier supporting the pedestrian overbridge is also near the highway boundary fence (Fig. 3).

11. In addition to the proximity of these features there was a need, about 3 years ago, to introduce a major maintenance contraflow at this site and with the flow of HGVs running on the hardshoulder it was decided to provide the extra protection. Although the safety fence has done 'its job', the design in appearance terms could be described as 'a little heavy'.

PART 2: ALTERNATIVES TO STRENGTHENING

Fig. 3. DS,DROBB Safety Fence on M6 at Corley Service Area.

CONCRETE SAFETY BARRIERS

12. <u>Permanent Normal Containment Concrete Safety Barriers (1.5t)</u>. Safety fences are designed to absorb some of the energy of a vehicle impact by elastic/plastic deformations as they deflect transversely by varying amounts depending upon the momentum of the impacting vehicle. Concrete safety barriers however, are intended to provide the level of containment without significant deflection or deformation by relying on the mass of the barrier.

13. <u>Permanent British Concrete Safety Barrier (BCB)</u>. Research on concrete barriers in the early 1970's used the British Concrete Barrier (BCB) which has a profile fairly similar to the 'New Jersey Barrier' used extensively in North America and on the Continent (Fig. 4). Unfortunately, impact tests with small European type cars like the Mini demonstrated that the BCB was unacceptable with high speed impacts, as the sloping profile tended to induce such vehicles to climb up the barrier and then roll-over. For this reason the BCB design has only been allowed on UK roads which have been restricted to a speed limit of 50mph (80km/h) or less.

14. A 16.5t HGV travelling at 80km/h and impacting at 15° was satisfactorily contained and redirected, but the vehicle rolled some 31° towards the barrier due to the much higher centre of gravity of the loaded vehicle. However, in a similar test with a 39.2t articulated HGV the vehicle actually breached the barrier and then travelled along straddling its top.

258

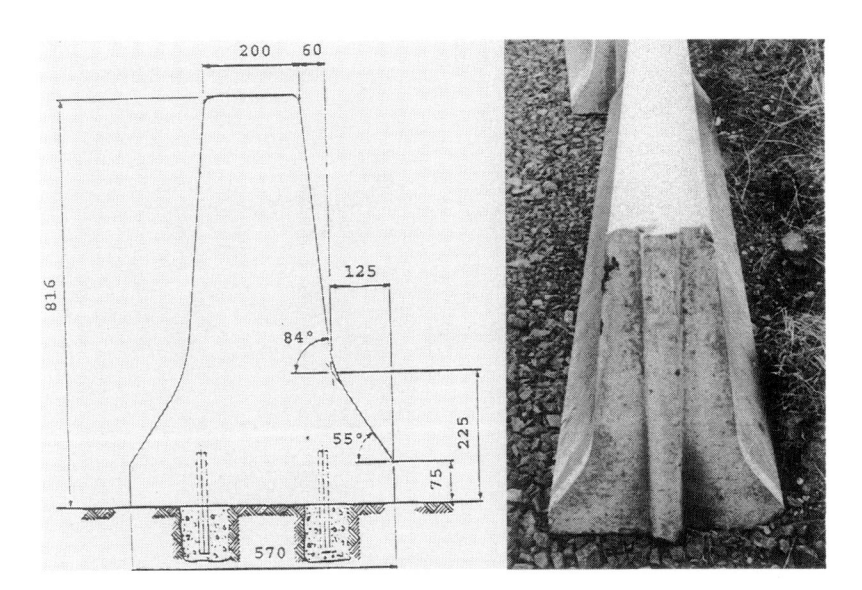

Fig. 4. Permanent British Concrete Safety Barrier (BCB)

15. Permanent Vertical Concrete Safety Barrier (VCB - 1.5t). As a consequence the BCB has, with the co-operation of the British Cement Association and Industry, been redesigned to have a near vertical profile as shown in Fig. 5. Dynamic vehicle impact testing with small and medium sized cars as well as 16t vehicles, has demonstrated that this vertical concrete safety barrier (VCB) is suitable for deployment on 70 mph (113km/h) roads. The change in profile reduced the roll of cars to a very low level and this design of barrier is now being installed on UK roads, particularly where cross-section widths are a design constraint.

16. The normal containment VCB is suitable for installation in central reserves where the available width is not sufficient to accommodate a metal safety fence plus the associated deflection requirement. Examples of the VCB can already be seen on the A12 at Gorleston & A46 south of Leicester (Fig. 5), M20 between Junctions 5 to 6, A3 South of Guildford at the junction with the A31 (Hogsback), A167 in Durham, Brechin By-pass in Scotland and in the next few months on the M25 between Jcts 10 to 11 and also Jcts 7 to 8.

17. Although the VCB is still only designed to contain a 1.5t car at 113km/h impacting at 20° it is expected, due to the lack of barrier deformation, to safely contain vehicles of much greater weight depending upon the approach angle of the errant vehicle. Design details of the VCB both for pre-cast units and slipformed construction are now given in the Highway Construction Details and they will also be included in British Standard, BS 6579:Part 9 in the near future (ref.3).

PART 2: ALTERNATIVES TO STRENGTHENING

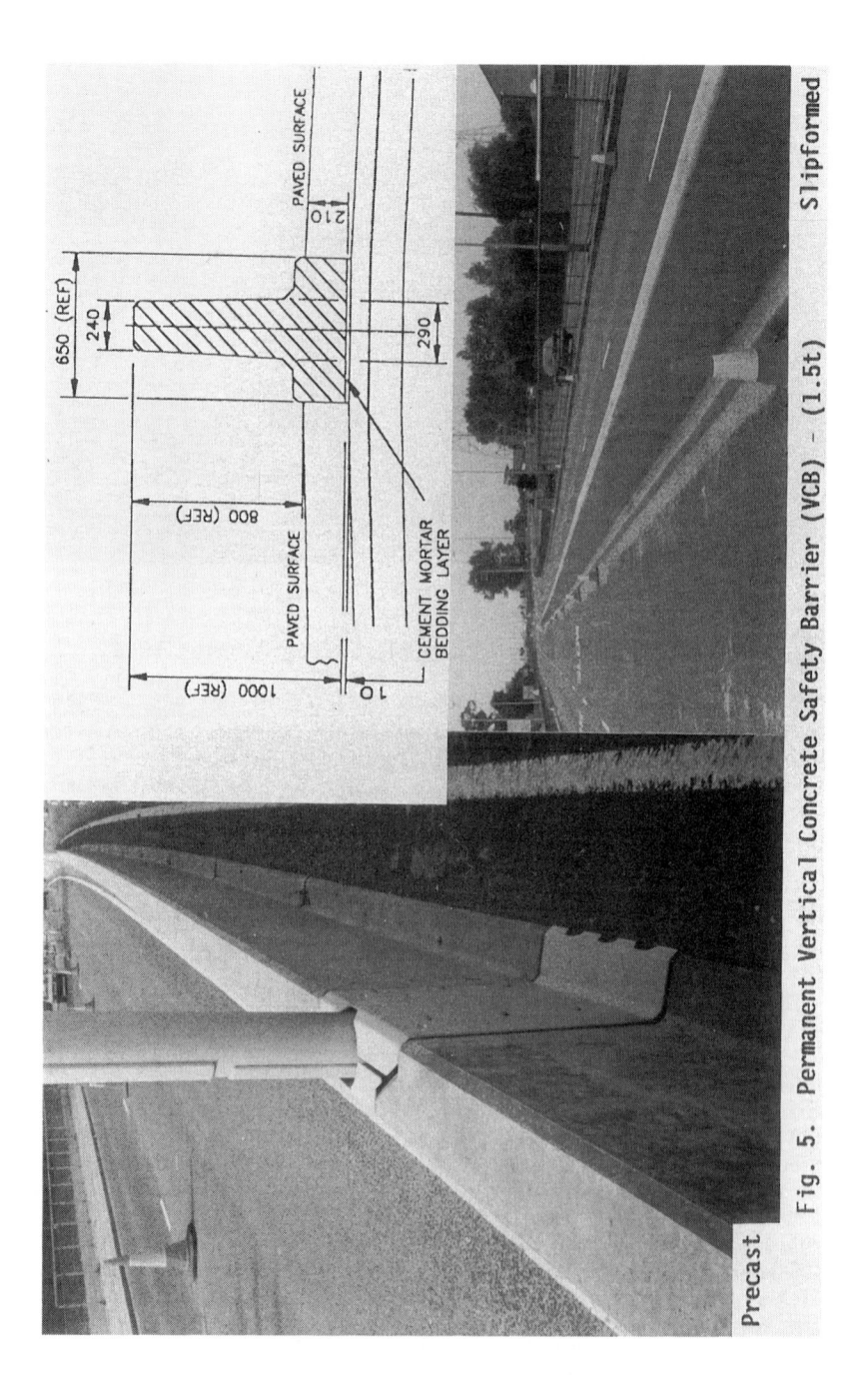

Fig. 5. Permanent Vertical Concrete Safety Barrier (VCB) - (1.5t)

Fig. 6. Temporary Vertical Concrete Safety Barrier (TVCB).

18. <u>Temporary Vertical Concrete Safety Barrier (TVCB - 1.5t).</u> About 6 years ago during the course of a motorway widening scheme, it was found to be beneficial, as a safety measure, to provide temporary precast BCB profiled units as a protection to work areas and the associated workforce. Subsequently, these temporary units, which were joined together by means of a steel pin through interconnecting wire loops cast into the ends of the units, were dynamically tested with a 1.5t car at 50 mph (80km/h) at 20°.

19. Whilst the system contained the test vehicle, there was an excessive amount of lateral movement (1.0m) and the previously established phenomenon of the test vehicle tending to ride up the barrier occurred again. Such a potential encroachment into a 'work zone' was deemed unacceptable and following further R&D work based on the successful impact testing with the permanent VCB, a design of temporary vertical concrete safety barrier (TVCB), utilising bolted scarf joints, was successfully tested at 80km/h (Fig. 6). The TVCB not only deflected significantly less (i.e. only 150mm) but it redirected the test vehicle in a safe and satisfactory manner.

20. The TVCB, which can also be constructed using slipforming techniques as well as in the normal precast units, was first introduced for use at road work sites in 1989; since then their deployment has increased greatly thereby resulting in much improved and safer conditions for both the work zone and those inside it, and also for the general traffic on the highway. It is estimated there are currently about 25,000 TVCB units available for use on our roads. So far there does not appear to have been a serious casualty resulting from the use of the TVCBs.

PART 2: ALTERNATIVES TO STRENGTHENING

Higher Containment Vertical Concrete Safety Barriers (30t).

21. Permanent Higher Vertical Concrete Safety Barrier (HVCB). The Department, as part of its Motorway Widening Programme has, in the past 3 years, been assessing the potential for increasing the number of traffic lanes on the relevant sections of motorways. At the same time there has been the need to incorporate the revised enhanced collision loading requirements detailed in Departmental Standard BD48/93 (ref. 4). The implications of these requirements have generated a great deal of interest in the potential of concrete as an effective protective barrier in front of vulnerable structural supports because of its rigidity and minimal clearance requirements.

22. The combination of the above factors led to the examination of a 1.2m higher containment vertical concrete safety barrier (HVCB), to initially contain a 16t vehicle (Fig. 7). The HVCB has a near vertical profile similar to the 0.8m high normal containment VCB (Fig. 5). Again, as a joint venture with BCA and Industry, the opportunity to examine the possibility of slipforming the barrier was pursued, as this method of construction had the potential to be much more economical than precast units, particularly for significant lengths of safety barrier.

23. In July 1990, following various trials and investigations into concrete mix design and adjustments to the slipforming machine, a trial length of the HVCB was constructed at the Motor Industries Research Association's (MIRA) crash test site. Four impact tests using 16t and 22t rigid HGVs travelling at 80 km/h and impacting at 15° and 20° were successfully completed. In each test there was no significant damage or lateral movement of the HVCB and in all cases the test vehicle was contained.

24. A final impact test on this trial length of slipformed HVCB (Fig. 7) was carried out in late 1991 to establish whether it could satisfy the 'P6 level of containment' (30t x 64km/h x 20°). The 30t tanker was successfully contained, although it rolled on to the HVCB so that the bodywork of the tank encroached for a short time beyond the rear of the HVCB by just less than 1.0m (Fig. 8).

25. Examples of permanent HVCBs constructed using in-situ slipforming and by in-situ casting between fixed forms, can be seen on the M4 between Jcts. 4b to 5 and M6/M1 Catthorpe Interchange (Fig. 9), and at the Severn Bridge Toll Plaza. Numerous other HVCB installations are currently in hand, particularly on the two M25 widening schemes which have just commenced (i.e. Jcts 10 to 11 and Jcts 7 to 8). On these two schemes the HVCBs are being used in conjunction with normal containment VCBs to provide collision load protection at many of the overbridge supports, thereby overcoming the requirement for very costly and time consuming replacement of many of the structures and their supports. A further use of the HVCB in the future will be to provide protection to the supports of overhead cantilever and portal gantry signs (ref. 5).

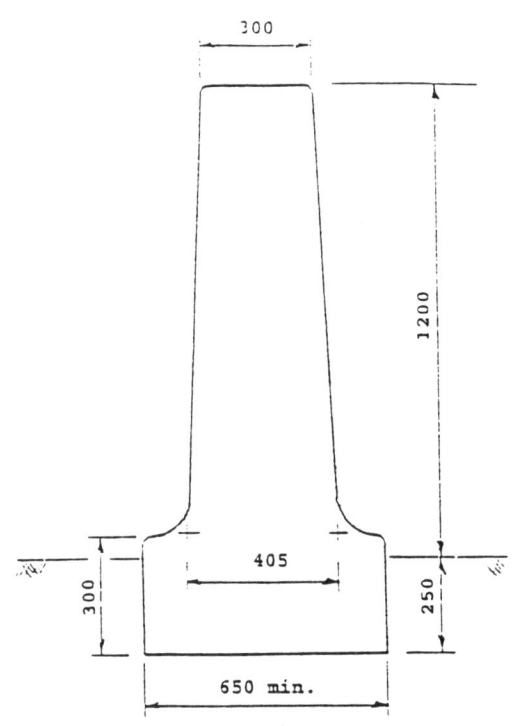

Fig. 7. Higher Containment Vertical Concrete Safety Barrier (HVCB) - 30t (4 axle rigid tanker) x 64km/h x 20°

PART 2: ALTERNATIVES TO STRENGTHENING

Fig. 8. 1.2m High Slipformed HVCB - 30t Tanker Impact at 64km/h and 20°

M4 Jcts 4b to 5

M4 Jcts 4b to 5

M1/M6 Catthorpe Interchange

Fig. 9 Permanent Higher Vertical Concrete Safety Barriers (HVCB)

PART 2: ALTERNATIVES TO STRENGTHENING

26. <u>Temporary Higher Vertical Concrete Safety Barrier (THVCB).</u> Consequent to the on-going inspection and assessment of bridge supports over recent years and the need to undertake essential repairs and strengthening works, an urgent requirement for a temporary higher containment restraint system has been highlighted to protect such structures and any temporary supports whilst the remedial works are carried out. Departmental Standard BD48/93 (ref. 4) requires any structure within 4.5m of the carriageway traffic to be able to resist both the 'Main load' and 'Residual load' collision components specified in the document.

27. In this type of situation, any safety barrier installation would be of a relatively short overall length and likely to be required only for a comparatively short period and therefore it should be easily removable and preferably re-usable. With these considerations in mind it was decided to build and impact test an extended version of the normal containment precast 'universal' design of VCB. This temporary higher containment barrier (THVCB) has a height of 1.4m, of which 200mm must be embedded in a hardened area of paving, although no specific foundations are needed. To give adequate stiffness across the bolted scarf joint, a 30mm thick shear plate is used and this is secured to each unit with four 20mm, Grade 8.8 bolts (Fig. 10).

28. <u>Lateral clearances for THVCB and HVCB to support members.</u> Although the details given in Figs. 11 to 14 refer specifically to the type of temporary VCB or HVCB required, relative to the lateral position of the system to any temporary propping and its ability to cater for 'main and residual load components', the same basic requirements apply for the provision of a HVCB in front of a permanent vulnerable structural support. The potential for using VCB or HVCB at structures with restricted headroom is shown at Fig. 15.

29. Full details of both the THVCB and the HVCB designs are currently being finalised and these will be issued as Highway Construction Details.

CONCLUSIONS

30. There are now systems available that are capable of protecting both permanent and temporary structural supports from the effects of HGV impacts. These systems have been dynamically tested and successfully installed at specific sites. They will play an increasing role in the 15 year bridge rehabilitation programme by enabling upgrading works to be carried out in a cost effective, safe and efficient manner whilst still allowing the maximum possible use of adjacent carriageway space for highway traffic. The deployment of the permanent higher vertical concrete safety barrier will also play an important role in reducing the cost and time associated with motorway widening works by providing protection to existing overbridges with vulnerable supports or profiled deck soffits.

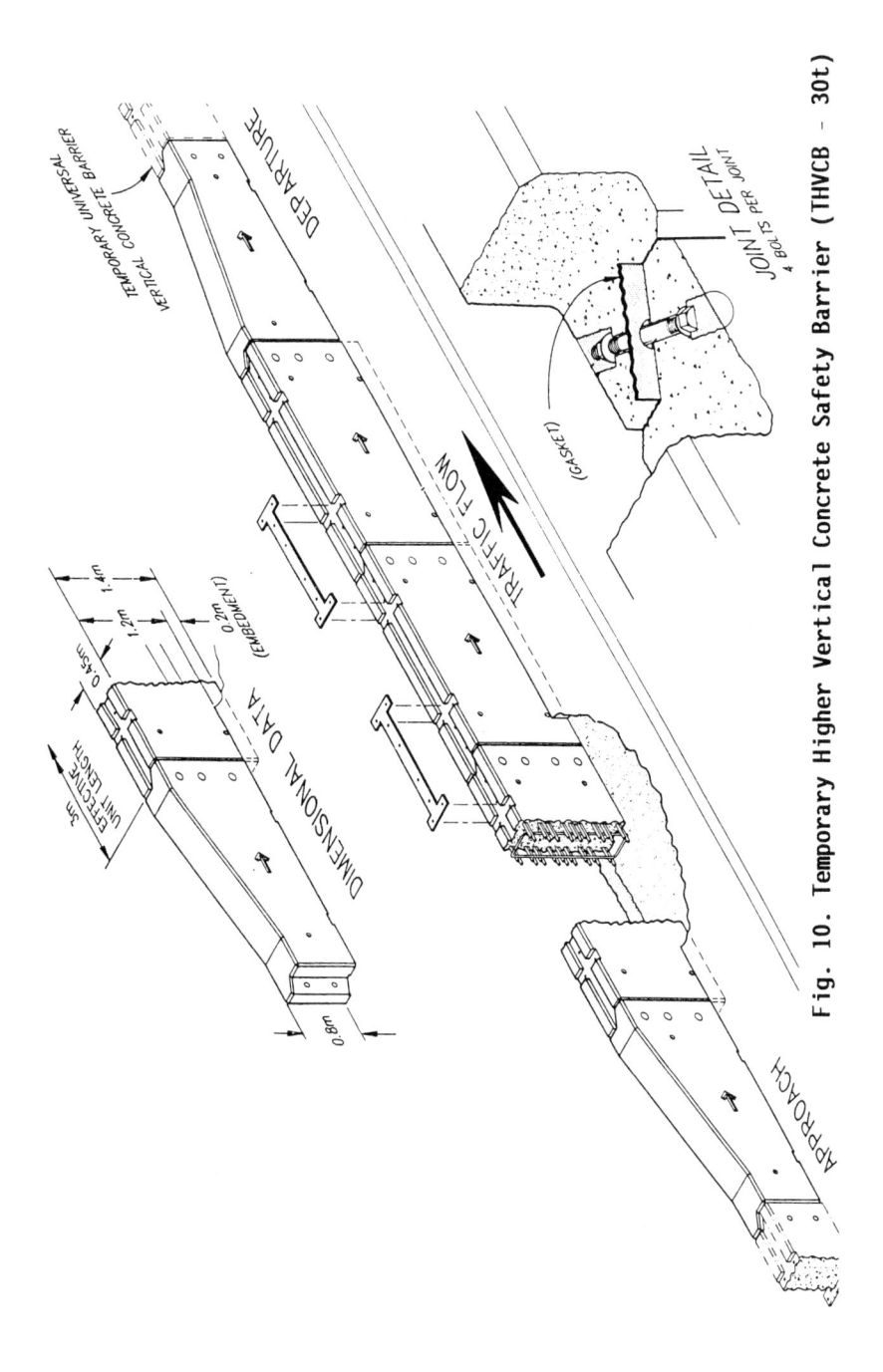

Fig. 10. Temporary Higher Vertical Concrete Safety Barrier (THVCB - 30t)

PART 2: ALTERNATIVES TO STRENGTHENING

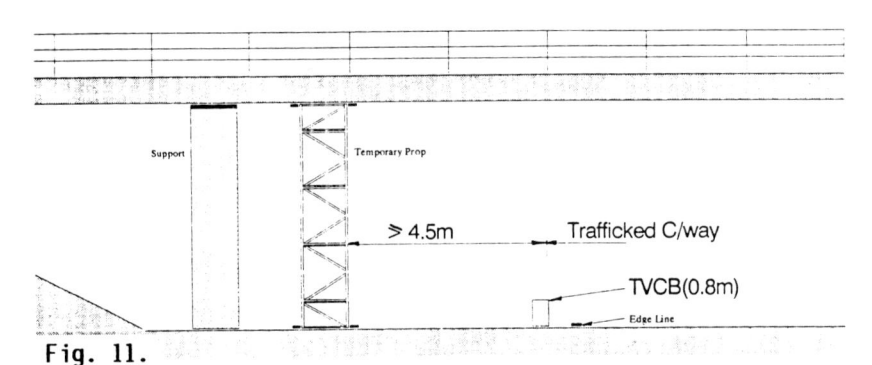

Fig. 11.
TEMPORARY SUPPORT 4.5M OR MORE FROM EDGE OF TRAFFICKED CARRIAGEWAY

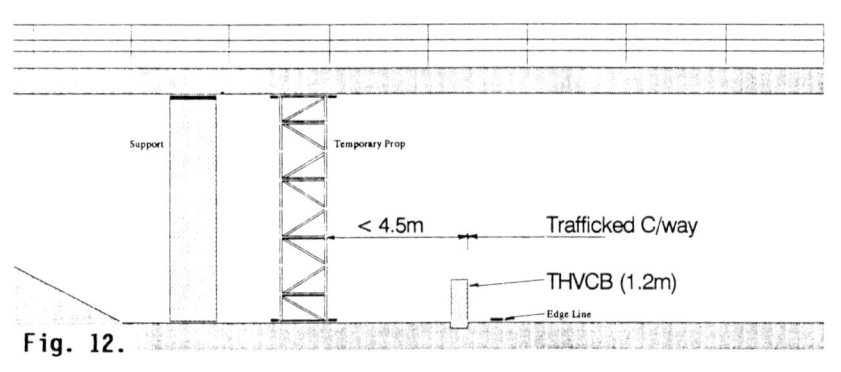

Fig. 12.
TEMPORARY SUPPORT LESS THAN 4.5M FROM EDGE OF TRAFFICKED CARRIAGEWAY

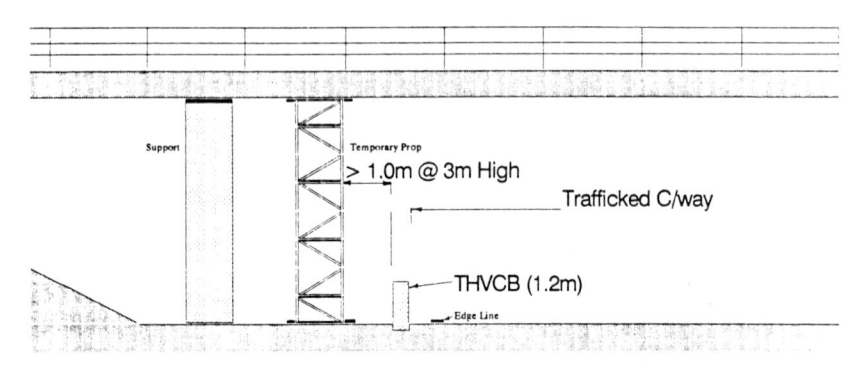

Fig. 13. TEMPORARY PROPPING – NO COLLISION LOADING REQUIREMENT

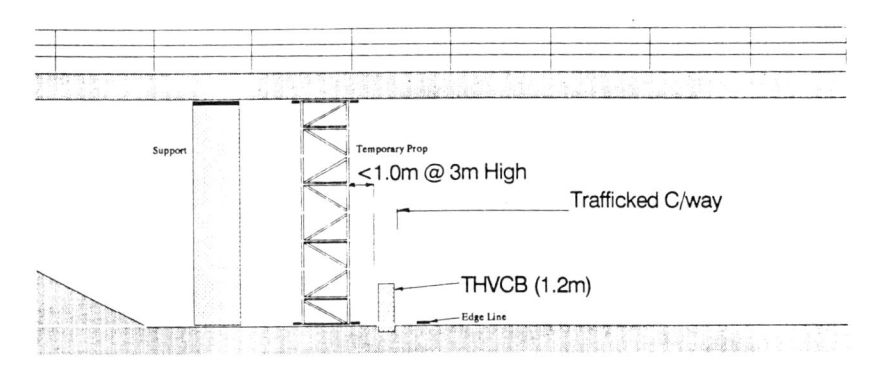

Fig. 14. TEMPORARY PROPPING – TO CATER FOR RESIDUAL LOADING

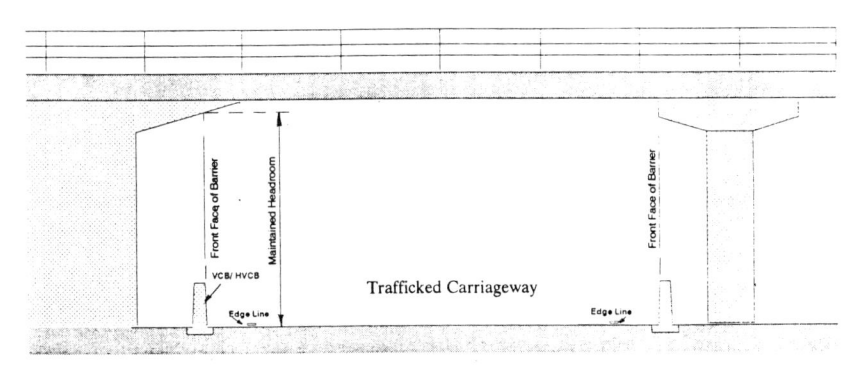

Fig. 15. VERTICAL CONCRETE SAFETY BARRIER AT RESTRICTED HEADROOM

ACKNOWLEDGEMENTS
This paper is published by permission of the Deputy Secretary, Highways Safety and Traffic, Department of Transport.

REFERENCES
1. Design Manual for Roads and Bridges, Departmental Standard BD 52/93, The Design of Highway Bridge Parapets, HMSO (DMRB 2.3), 1993.
2. Manual of Contract Documents for Highway Works, Volume 3: Highway Construction Details, HMSO (MCHW 3), 1991.
3. British Standard, BS 6579: Part 9 Specification for Permanent and Temporary Vertical Concrete Safety Barriers, BSi Standards (In preparation).
4. Design Manual for Roads and Bridges, Departmental Standard BD 48/93, The Assessment and Strengthening of Highway Bridge Supports, HMSO (DMRB 3.3), 1993.
5. Design Manual for Roads and Bridges, Departmental Standard BD 51/94, Design Criteria for Portal and Cantilever Sign/Signal Gantries (In Draft).

© Crown Copyright. The views expressed in this paper are not necessarily those of the Department of Transport.

Is your strengthening really necessary?

D. W. CULLINGTON, Bridges and Ground Engineering, Transport Research Laboratory, and C. BEALES, Research Programme Co-ordinator, Department of Transport Highways Engineering Division

SYNOPSIS. The Transport Research Laboratory (TRL) has carried out tests on bridges, bridge components and models to identify reserves of strength that could be made use of in assessment. Reserves have been found in a number of masonry arch and short-span non-arch bridges, and some specific details. In some cases the results have led to revised methods of assessment: in others to data or advice that must be used with judgement by the engineer. On occasions in-situ load tests may be used to augment the assessment process.

INTRODUCTION

1. The bridge assessment and rehabilitation programme has revealed a substantial number of bridges to be inadequate for further unrestricted service: that is to say, the *assessment* process has led to the conclusion that weight limits or carriageway restrictions are necessary to avoid overloading. In the first instance the assessment of a bridge is normally carried out using design standards, or simple assessment methods. If this does not indicate the desired capacity, various steps may be considered before a decision is taken to strengthen or replace a bridge.

2. The Department of Transport (DOT) programme of research into bridge assessment has included the testing by TRL of a number of bridges, bridge components and models. In many instances it has been possible to demonstrate the presence of useful reserves of strength: but this will not always be the case, and simply observing that a bridge is showing no signs of distress in service is not in itself sufficient. Adequate margins of strength must be demonstrated; and although some bridges have reserves of strength above their assessed capacity this cannot be taken for granted.

RAISING THE ASSESSED CAPACITY

3. Engineers have become familiar with a number of techniques that may demonstrate a higher carrying capacity than obtained initially from calculations. These include:

   (a) The use of measured material properties in combination with reduced material factors;

(b) Adopting an alternative theoretical model, for example yield-line analysis, compressive membrane action, or non-linear analysis;
(c) In the case of a non-conforming detail or simplifying assumption, a departure from standard may be sought when there is evidence that the code requirement is conservative;
(d) If there are published data on the collapse testing of a similar bridge, it may be possible to use this as an example to validate an alternative theoretical model;
(e) If a detail or structural form cannot be reliably dealt with theoretically, a laboratory test may be considered;
(f) If the conditions seem right, an in-situ load test may be appropriate.

4. In all cases the relative costs of the alternatives and their probability of success need to be assessed.

MATERIAL TESTING

5. It is common knowledge that some advantage may be gained by material testing if it can be shown that the in-situ strength is greater than originally assumed. The DOT advice note BA 44/90 (Ref. 1) gives guidance on obtaining values of worst credible strength for concrete and the assessment standard BD 44/90, gives reduced material factors that may be used with them to give an even greater advantage.

6. Engineers seem to have been using this provision without undue difficulty, but the advice has to be adapted to suit the circumstances. By way of illustration, the following example of multiple core strength tests may be found illuminating. The bridge was a three-span reinforced concrete structure and was about to be demolished. It was thus possible to obtain a relatively large number of cores grouped together at particular locations (cross-sectional "regions").

7. In all, 101 cores were obtained, approximately one per 10m^3, of which 81 came from eight specified locations. Taken overall, the range of strengths was 36 to 101 N/mm^2. Most locations produced a significant range of strengths: for example 36 to 83 N/mm^2. Estimated "characteristic" strength values at single locations ranged from 45 to 72 N/mm^2, for which the corresponding values of worst credible strength "in a uniform area" (from the formula in BD 44/90) were 62 and 75 respectively and the means 65 and 79. The overall mean was 71, the characteristic 52 and the "worst credible" from the formula was 70 - all values being in N/mm^2.

8. These and other similar results suggest that areas of uniform quality can be very local. Moreover, if the formula in BA 44/90 is applied to a large structure or component the worst credible strength tends towards the mean value (actually a safe estimate of the mean using an assumed value for standard deviation) as the number of cores is increased.

PART 2: ALTERNATIVES TO STRENGTHENING

9. Supplied with all this information in an assessment the engineer could have used 36 N/mm^2 as a worst credible value for the structure initially. If this was not sufficient, a higher concrete strength could be taken at critical locations. Without as many core results the situation would have been less certain. It was found that as few as eight cores spread along the length of the structure would have given a reasonable estimate of the overall mean in most cases but would have overestimated the characteristic value by more than 20% for one in ten samples. To achieve a safe but not over-conservative assessment the engineer might need a sensitivity study, pay attention to critical areas and apply judgement in the selection of the number and locations of cores.

10. For flexural capacity, the steel strength is often the governing factor. Here the chief difficulty lies in the size and number of specimens required. Small specimens tested according to BS EN 10 002 (Ref. 2) can give different values from BS 4449, but the latter document requires long specimens for testing which is inconvenient. BS 4449 (Ref. 3) gives guidance on the use of small samples to obtain the characteristic strength. However, if a full-width failure is under consideration, a value for steel strength closer to a safe estimate of the mean value might be more appropriate.

PROBLEM DETAILS AND LABORATORY TESTING

11. Some problem details that have come to our notice at TRL include the absence of shear links in prestressed beams, interfacial reinforcement spaced too widely, non-complying configurations in filler-beam decks, inadequate anchorage in and the treatment of webs in slabs with circular voids. Selective laboratory testing has thrown some light on these areas. The DOT standard BD 44/90 relaxed the requirements for shear links (Ref. 4) and the advice note BA 44/90 refers to tests on beams with inadequate interfacial shear reinforcement. In the latter instance, a short series of load tests to failure on 9m-long prestressed beams gave satisfactory results even when the spacing and amount of interfacial reinforcement fell outside normal limits. Future testing may throw further light on these areas before the assessment programme is complete.

Half-scale bridge tests

12. Tests at TRL on two half-scale models of bridges of "modern" design have indicated the presence of reserves of strength in some details; but they have also shown that reserves of strength in global modes of failure cannot be guaranteed in such bridges.

13. The first (see Fig. 1) consisted of a pre-tensioned beam and in-situ slab deck having a 9 m simply-supported span, loaded by highway types HA and HB applied in combination. For the collapse, HA load at the ULS factored level was applied in one lane while in the other lane, load in the HB configuration was applied in increments up to and beyond the maximum load.

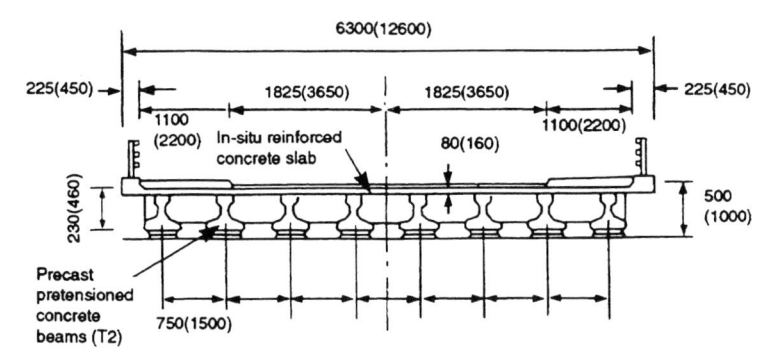

Fig. 1. Cross-section of half-scale model (full-scale dimensions in brackets).

14. Failure occurred at a load factor λ of 3.17 (where λ refers to multiples of the ULS HB load) initiated by shear in the prestressed beams under the HB axles. This apparently high value of failure load is partially explained by material safety factors and characteristic (rather than mean) strengths used in the design. Once allowance has been made for these, shear failure was indicated by calculation at a λ value of 2.8. For some "details", however, the live load effect present at maximum load was around twice the predicted strength - notably for slab bending, interfacial shear, web crushing, and torsion/shear near the supports. Discounting these secondary effects as unlikely to result in failure (Ref. 5) the model failed in the expected primary mode. The difference between λ = 2.8 and 3.17 indicates a modest margin of strength (13 per cent) above the calculated capacity obtained from an elastic grillage analysis and unfactored design code methods.

15. It should be noted, however, that in the secondary areas mentioned there were reserves of strength that might reasonably be accounted for in assessment. No specific rules can be offered from the results of a single test - but for similar structures a comparison with the test behaviour might be used to advantage. For a large structure, or several similar structures with the same detail, an investigation by laboratory testing could be rewarding.

16. A second model test was carried out on a steel beam and slab (composite) deck with similar overall dimensions and loading configuration to the concrete bridge. Failure occurred in longitudinal flexure when the load reached a λ value of 2.2. This was 8 per cent more than the calculated capacity obtained from an elastic grillage analysis using measured mean strengths for the materials and unfactored design code methods. In the test, the mode of failure was more complex than indicated by the calculations, being initiated by crushing in the parapet upstand. However, in broad terms, there was little overall reserve of strength as was the case for the concrete deck.

PART 2: ALTERNATIVES TO STRENGTHENING

17. It is concluded that for modern bridges of simple beam and slab configuration, it should not be assumed that reserves of strength are always present in primary failure modes.

FULL-SCALE TESTS TO COLLAPSE

18. A number of collapse tests have been carried out by TRL on redundant bridges. In the case of arch bridges there is a substantial population of similar structures and a reasonably large number available for collapse testing. These tests have been used for the calibration of revised assessment methods.

19. For other structural types the population is smaller and therefore economics preclude a long testing programme. Moreover, the number of redundant bridges available for testing is small, and variations in the precise structural form are more significant. Full-scale collapse testing for these structures can more readily provide examples for back analysis rather than new rules, which by their nature have to be well established and suitable for general application. Most of the non-arch bridges tested have been older types, of short span and without discrete bearings. Where reserves of strength have been indicated, they have been attributed mainly to actions and restraints commonly ignored in assessment.

Masonry arches

20. Eight collapse tests have been carried out on masonry arch bridges having spans of 4.9 to 18.3m (Ref. 6) and two large-scale model have been tested. In the tests, transverse line loads were applied at a longitudinal position calculated to give minimum strength - usually at the quarter point. Loads were in general applied through jacks reacting against ground anchors. Maximum loads were between 228kN and 5600kN, most of the failures occurring by the formation of a 4-hinge mechanism.

21. The 'MEXE' method described in BD 21 (Ref. 7) is a commonly-used simple method for assessing masonry arch bridges from which allowable axle loads are derived. Comparision between the maximum test loads and the allowable axle loads is not straightforward. However, for the bridges tested, it may be concluded that the MEXE assessment is generally satisfactory for the ultimate limit state and may be consevative in some cases by a factor up to three. The results of the tests are quoted in the advice note BA 16 (Ref. 7) together with the results of calculations carried out according to a variety of methods of assessment.

Jack-arch/hogging plate bridges

22. Three bridges of this type have been tested: two comprising brick jack arches spanning transversely between steel girders and one with steel 'hogging plates' and steel girders. Assessments are normally carried out using the simple distribution analysis given in BA 16. Jack arches are assumed to spread the lane load between a number of beams, the resistance of each beam being taken as that of the beam alone acting as

simply supported with no further assistance from the arch or fill

23. It was found in the tests that the distribution between beams obtained from BA 16 was reasonable, although sometimes conservative particularly for the edge beams. At low loads the measured stiffness of the deck was generally several times that calculated for the bare beams, and at the ultimate limit state there was a residual effect that increased the capacity to a useful extent. This could be accounted for by composite action between the beams and the rest of the deck, with a possible contribution from arching to the supports.

24. Jack Arch Bridge 1 (Fig. 2) had two 3.5m spans each having 5 rolled steel joists and 2-ring brick arches. The bridge was assessed and found to have a reduced loading capacity of 7.5 tonnes and was to be replaced. Tests were carried out on various parts of the structure, using jacks reacting against ground anchors. At low loads the deck was up to five times as stiff as the bare beams. A maximum load of 1500kN was applied on a knife edge across the middle of one of the spans before one of the steel beams fractured. The second span resisted a more concentrated load of 1350kN applied at four points to simulate a vehicle bogie. The response of the structure was consistent with simply-supported behaviour and composite action between the beams arches and fill, leading to a resistance of the deck of twice that of the calculated unfactored live load capacity. Transversely the jack arches remained able to distribute load between beams beyond the maximum load.

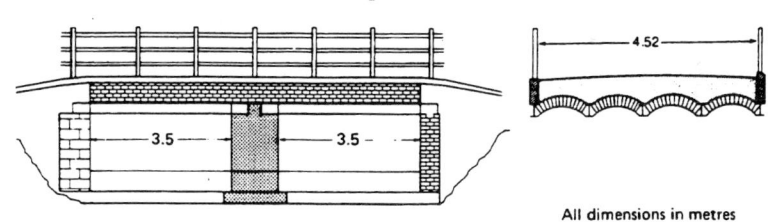

Fig. 2. Jack arch bridge 1.

25. Jack Arch Bridge 2 was of similar construction but used steel plate girders spanning 8m onto brick abutments. A similar series of tests was carried out. Load/strain data indicated that the deck had greater stiffness than the girders alone and that there was a degree of end-fixity at the supports. A maximum load of 2100 kN was applied to a simulated bogie at mid-span (at which point the ground anchors failed prematurely) without any observed damage to the integrity of the structure.

26. The hogging plate deck (Fig. 3) had a single 4.8m span with six steel beams. Weak concrete had been placed over the hogging plates up to the level of the top of the beams. Above this was approximately 250mm of fill and surfacing. Single

PART 2: ALTERNATIVES TO STRENGTHENING

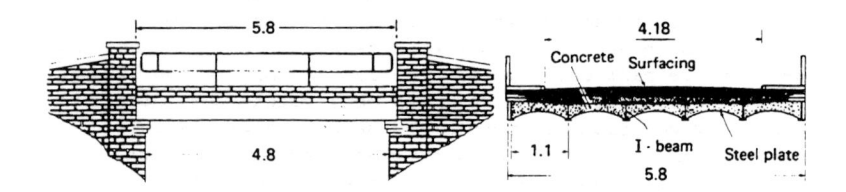

Fig. 3. Hogging plate bridge

point loads were applied to the road surface using a jack reacting against kentledge on a frame over-spanning the bridge.

27. Failure loads were approximately 810kN for the hogging plate, 740kN for an internal beam and 400kN for an edge beam. In the tests at low load the bridge was up to four times as stiff as that indicated by the bare beams alone, and in the failure tests about twice as strong (Fig. 4). The tests confirmed that the edge beam is a relatively weak area in bridges of this type and improved capacity may be obtained by restricting vehicle access in this area.

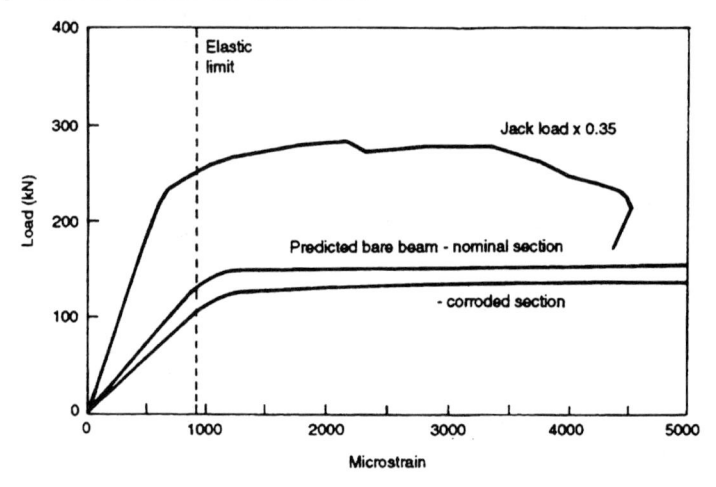

Fig. 4. Load-strain for load between beams on hogging plate deck

Filler beam bridge

28. One such bridge was tested to collapse. It had been constructed using 305mm deep RSJ sections as filler beams spaced at 457mm centres and cast within a 380mm thick concrete slab. There were no tie bars between the beams or transverse reinforcement within the slab. The beams spanned 5.8m and the deck was cast directly onto the abutments.

29. A structural assessment indicated that the bridge had inadequate longitudinal capacity in flexure unless adequate transverse distribution could be assured: but as the transverse

flexural capacity was very low this was not possible. For the purposes of the tests, a critical load case was calculated to be when two heavy goods vehicles passed on the bridge - this form of loading accentuating the transverse load effects. Two tests were carried out, the first with two point loads spaced 700mm laterally and the second with four point loads spaced at 700mm x 1800mm, both representing adjacent wheel loading.

30. In the first test cracking commenced at 1000kN and continued to develop up to 1900kN when the test was stopped, the failure load being estimated as 2200kN by extrapolation. In the second test cracking commenced at 1060kN, the failure mechanism started to develop at 2500kN and maximum load was reached at 2900kN. The mode of failure was flexural and ductile.

31. An assessment based on a simple dispersal of load in the transverse direction indicated capacities of about one fifth of the failure loads obtained for the two tests. Back analysis was able to explain the high test loads by assuming composite action, longitudinal end restraint and transverse distribution by a raking strut and torsion model.

32. Low and Ricketts (Ref. 8) conclude from the favourable distribution characteristics found in the tests that it is safe to assess these decks using an elastic load distribution with the deck modelled as a 'shear key deck' with uncracked section properties.

Reinforced concrete beam and slab bridge

33. The bridge tested was a 16-span beam and slab structure, cast in-situ having a total length of 230m. A typical span is shown in Fig. 5. An inspection showed that there was severe corrosion in the main beams and the assessed loading was reduced to 17 tonnes. In view of the deterioration it was decided to replace the bridge and the opportunity was taken to test an 11.8m continuous span to collapse.

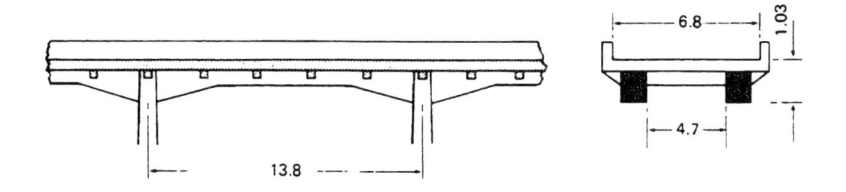

Fig. 5. Concrete beam and slab bridge

34. Four tests were carried out (Ref. 9): on a cross-beam loaded centrally, on a slab using a 1200mm line load and on two main beams loaded at 4.6m from the support. The beam failed as a beam-slab combination by shear and punching, the deck slab by punching after initially cracking in yield-line fashion and the main beams cracked but resisted failure.

PART 2: ALTERNATIVES TO STRENGTHENING

35. The two main beams behaved in a similar manner despite visual evidence of more serious deterioration in the north one. Flexural, then shear cracks formed, and there was cracking and spalling in the hogging regions over the piers. The tests were terminated when strands of the ground anchors began to fail. The load / displacement curve for a gauge under the load position is shown in Fig. 6. It indicates that the beam capacity was unlikely to have exceeded 6000kN.

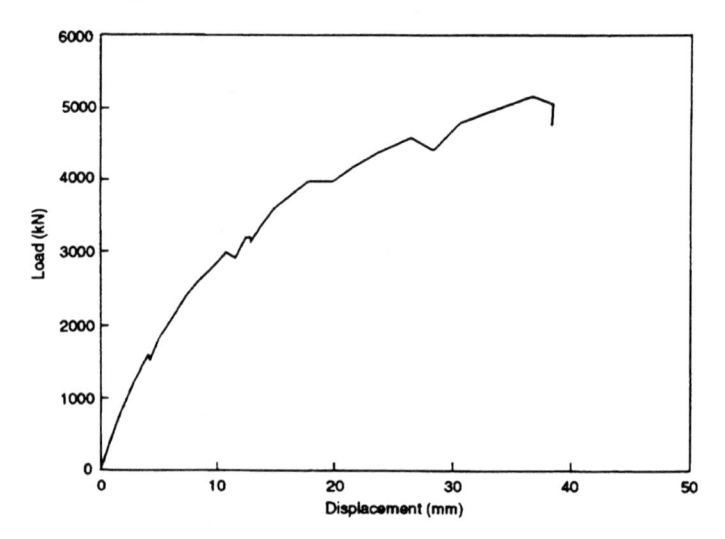

Fig. 6. Load-displacement for Test 3 on main beam of beam and slab bridge

36. Calculated failure loads using conventional analyses tests indicated that the cross-beam would fail at 750kN (cracking 930, failure 2480kN), the deck slab would fail at 1250kN (cracking 600, failure 2910kN), and the main beams would fail at 1130kN (cracking 1000, carried 5170 and 5310kN). The increase in strength was attributed to end restraints on members, structural interaction between elements and reserves of reinforcement strength. Back analysis using strut and tie models and compressive membrane action was able to explain the observed maximum loads satisfactorily.

Reinforced concrete slab structures

37. Two structures of this type have been tested. The first consisted of a single span simply-supported slab cast directly on mass concrete abutments: the second was an underpass in the form of a slab with pin connections at the supports restraining reinforced concrete abutment walls by propping action.

38. The simply-supported slab bridge was replaced following an assessment which indicated a safe capacity of only 7.5 tonnes, the governing effect being flexure in the transverse and longitudinal directions, with deficiency in steel

being the limiting factor. The dimensions of the bridge were: overall width 7.1m, clear span 4.7m and average slab depth 410mm.

39. The bridge was tested to collapse using a load configuration representing two adjacent twin-axle bogies. Failure was ductile and flexural, a maximum load of 3670kN being attained. The tests results indicated that the capacity of the deck was comfortably sufficient for full assessment loading. End restraints were present (moment and longitudinal force) which reduced the deflection of the deck under working loads and increased the carrying capacity.

40. The problem with the underpass was inadequate anchorage at the supports, leading to an insufficient assessed shear capacity in that area. The structure consisted of a 300mm slab providing a clear opening of 4.27m for farm access. A collapse test carried out on a 1m strip of the deck, at a shear span of three times the depth, produced a flexural failure at a load of 440kN, considerably in excess of the calculated shear-failure load ignoring the contribution from the reinforcement. Further laboratory tests are needed to investigate the behaviour at shorter shear spans.

LOAD TESTING FOR ASSESSMENT

41. In-situ load testing for assessment purposes is permitted in the DOT standard BD 21 and is the subject of a recent draft DOT advice note. This describes two forms of load testing. One is recognisable as the type described briefly in BD 21, referred to as supplementary load testing, in which the results are used to derive an alternative theoretical model of the bridge. The other is a proving test, which in its basic form would entail the application of loads equivalent to, and representative of, the required assessment load factored for the ultimate limit state. Although the advice note discusses this method, it does not recommend its use unless research can prove that it is acceptable.

42. The main obstacle to supplementary testing is the necessity to carry out most assessments at the ultimate limit state. In devising an alternative theoretical model for the bridge, means must be found to allow for changes in behaviour as the maximum load is approached - for instance, there may be a reduction in the contribution from the composite action of beams with the rest of the deck. This problem can be overcome in proving load tests, but the approach carries a risk that the structure will be overloaded during the test and suffer damage as a result. The draft advice note recognises the benefit of using data from previous collapse tests for comparison purposes when attempting to use a supplementary load test to devise a new assessment model for a particular bridge.

43. At TRL we have carried out a number of load tests of the supplementary type: some for research purposes on bridges which were then generally tested to failure and others for bridge

PART 2: ALTERNATIVES TO STRENGTHENING

owners requiring to retain their bridges in service. Examples are given below.

Jack-arch/hogging plate bridges

44. When the main beams are cast iron the assessment is based on permissible stresses, and therefore the application of supplementary load testing is direct, requiring little extrapolation. The aim is to demonstrate that the stresses occurring under the nominal assessment loads are less than those obtained from the original theoretical calculations. Reasons may include improved composite action with the deck materials, more favourable distribution between beams and restraint at the supports.

45. One bridge tested consisted of two 6.9m spans, each of which contained four cast iron beams supporting brick jack arches. It was tested using the rear bogie of an articulated trailer loaded with concrete blocks (Ref. 10).

46. The assessment standard BD 21 allows the section modulus for live loading to be increased by a factor D/d (maximum value 2.0) where D is the overall depth of the deck (less 75mm for surfacing) and d is the depth of the bare girder at mid-span. The calculated value for this bridge was 1.6 but the measurements indicated that a higher value of 2.6 or above would have been more appropriate. The simple distribution method indicated zero live load capacity when used with a composite action factor of 1.6 increasing to 3 tonnes capacity with a factor of 2.6. The benefit was fairly small on this bridge because of the high dead load to live load ratio.

47. A second bridge tested was similar in construction but wider, having 11 beams and a single span of 9m at a 29^0 skew. An assessment to BD 21 indicated a capacity of 7.5 tonnes (and Group 1 FE). The bridge was loaded by two articulated trailers with 2-axle bogies, each loaded to 190kN. Over 100 load cases were applied during a 12-hour closure. The tests indicated that the longitudinal soffit-strains and deflections were considerably less than calculated values. A capacity of 25 tonnes was arrived at by using the test data in conjunction with the load distribution method of BA 16, and confirmed using a grillage analysis with modified deck properties obtained from the test data.

48. When the beams are of steel rather than cast iron, assessment is required at the ultimate limit state. Tests on a hogging-plate deck were carried out using point loads from which proportion factors for single-axle loading were derived. The values obtained were in reasonable agreement with, but somewhat lower than, those obtained using the simple distribution method given in the advice note. However, for edge beams, the tests showed the method can be conservative, particularly when the traffic lane is some distance from the edge of the bridge.

49. For Jack Arch Bridge 1, tested to collapse, elastic tests using a loading trailer produced proportion factors that agreed reasonable well with those obtained from loading with a

single jack, both being rather less than those obtained from
the advice note. On the basis of an improved distribution, the
assessed capacity increased from 7.5 to 17 tonnes. Allowing
for an increase in flexural capacity based on the collapse test
results would have raised this to full assessment loading. For
Jack Arch Bridge 2, also tested to collapse, the measured
strains demonstrated that the distribution of load between the
girders was more favourable than that obtained from the simple
distribution method, proportion factors being 25% less.

Trough deck
50. Tests were carried out on a canal bridge carrying four
lanes of traffic and consisting of longitudinal steel troughing
supported by brick abutments (skew 38°, skew span 9.06m). An
assessment to BD 21 indicated that a weight restriction of 7.5
tonnes was required, one factor being a wide variation in fill
over the troughs. The bridge was loaded by two articulated
trailers. Strains measured on the soffit of the troughs at
mid-span were considerably lower than calculated, see Fig. 7.
The reduced structural response was attributed to load distribution, end fixity at the abutments, skew and composite
action.

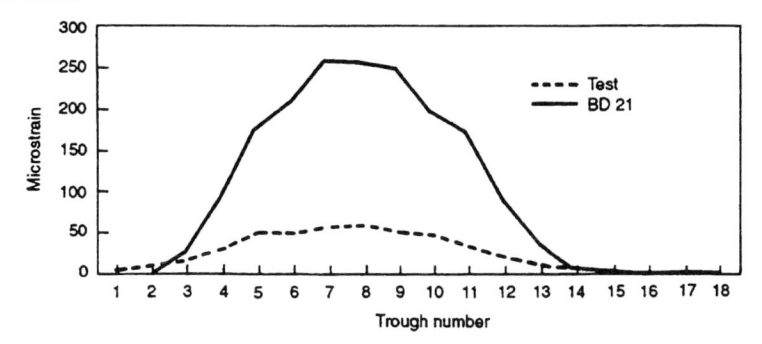

Fig. 7. Measured and calculated strains for
two-vehicle test on trough deck

51. The analytical method adopted was to calculate strains
for the test vehicle using the BD 21 method and compare them
with measured strains, thereby finding a reduction factor for
each trough. For the purposes of assessing the ultimate
strength it was assumed that end fixity and composite action
would reduce as the limit state was approached. The revised
model was then analysed using the BD 21 method and assessment
loading. Making conservative assumptions, a significant
increase in live load capacity was indicated.
52. Laboratory tests were carried out on a cast-iron floor
plate retrieved from a covered way complete with the original
cover, some 550mm deep. Static and cyclic load tests were
carried out under working load levels and finally the maximum
load was obtained in a test to collapse. BD 21 does not permit

PART 2: ALTERNATIVES TO STRENGTHENING

the use of the D/d stiffness factor in the assessment of this form of cast-iron deck; but the tests showed that it may be present, and an in-situ load test to evaluate it could be carried out safely.

Filler beam bridges

53. Vehicle load tests have been carried out on two filler beam bridges similar to the one tested to collapse. Both were constructed with 10 No. 8" x 6" RSJs within a 300mm (approx) concrete slab, one having a clear span of 4.25m and the other a 4.8m skew span at 39^0. Loading was by an articulated trailer, strengthened and modified to carry up to 200kN on each axle of its two-axle bogie. When fully loaded, this produced lane moments similar to those of HA loading.

54. The pattern of strains and displacements recorded was consistent to the results of the collapse test and the shear-key behaviour used to analyse it. A provisional assessment under the loading of BD 21 using the findings of these tests suggested that neither bridge required weight restrictions.

Mass concrete arches

55. Two bridges had failed assessment under HB loading of between 25 - 32 units. The bridge spans were approximately 15m with an arch 600mm thick at the crown increasing to 1200mm at the springings. The arches were constructed in approximately 5m sections with longitudinal construction joints.

56. Load tests were carried out using the DOT single axle 'HB' trailer, loaded in increments to 450kN. A series of longitudinal influence lines of strain and displacement were obtained by moving the trailer across the bridge at 1/8 span points. The results were interpreted by comparing them with the predictions of a plane frame linear elastic model, the arch being hinged at the crown and springings. Parameters of the load model were adjusted to match the test results.

57. By considering the form of various transverse distributions of strain, an idealised, general distribution plot was drawn, chosen to be conservative. Using this distribution the ULS thrust zone was found to be within the concrete arch everywhere. The arches were therefore assessed as being adequate for 45 units of HB. The test results also showed evidence of additional stiffness which was not invoked in the analysis.

Reinforced concrete slab

58. The reinforced concrete slab bridge cast on abutments that was tested to collapse was also tested before failure using two trailers each loaded on their rear bogies to 40 tonnes. In spite of being assessed for 7.5 tonnes, this bridge carried the 80 tonnes bogie loading at mid-span without signs of distress. Subsequent collapse test results indicated that had loading been increased to ULS factored assessment loading, some cracking might have occurred on the soffit.

THE WAY AHEAD

59. Before the current bridge rehabilitation programme is complete, action will have to be taken on a number of bridges that have so far been assessed as sufficient for reduced loading only. In some cases replacement or strengthening will be appropriate: in others, recourse to in-situ testing or one of the other alternatives may be considered. One of the consequences of the draft advice note on load testing will be to make engineers more conscious of the need to preserve the integrity of the structure during the tests and in later service. This may restrict the use of load testing in some cases.

60. Proponents of proving load tests will probably expect to provide evidence of such tests being safe and able to indicate sufficient capacity in some bridges. This is likely to be restricted to particular types of structure where helpful restraints are present (and may lead to an increase in capacity) but are difficult to quantify without a test.

61. There is also interest in applying a reliability approach to assessment, monitoring and load testing. In the case of load testing, this might allow a useful increase in assessed capacity without the application of unduly high loads. All forms of load test are to be welcomed if they extend the life of existing structures safely and economically.

ACKNOWLEDGEMENTS

62. Much of the work described in this paper forms part of the TRL programme of research for the DOT and is published by permission of the Chief Executive. The views expressed are not necesarily those of TRL or DOT.

REFERENCES

1. DEPARTMENT of TRANSPORT. The assessment of concrete bridges and structures. Departmental Standard BD 44/90 and Advice Note BA 44/90. Department of Transport, 1991.
2. BRITISH STANDARDS INSTITUTION. Tensile testing of metallic materials. BS EN 10 002, BSI, 1990, London.
3. BRITISH STANDARDS INSTITUTION. Carbon steel bars for the reinforcement of concrete. BS 4449, BSI, 1988, London.
4. CULLINGTON D.W. and RAGGETT S.J. Shear strength of some 30-year old prestressed beams without links. The Transport Research Laboratory, Crowthorne, 1991, Report RR327.
5. DALY A.F and CULLINGTON D.W. Assessment implications from tests on a model concrete beam and slab bridge. The Transport Research Laboratory, Crowthorne, 1991, Report RR309.
6. PAGE J. Masonry arch bridges. Proc 5th North American Masonry Conference. Urbana - Champaign, 1990.
7. DEPARTMENT of TRANSPORT et al. The assessment of highway bridges and structures. DOT, Scot. Off. Ind. Dep, Welsh Office, DoE N.Ireland, 1991, Standard and Advice Note BD 21 and BA 16.
8. LOW A McC and RICKETTS N.J. The assessment of filler beam bridge decks without transverse reinforcement. The Transport Research Laboratory, Crowthorne, 1993, Report RRS383.

PART 2: ALTERNATIVES TO STRENGTHENING

9. RICKETTS N.J. and LOW A. McC. Load tests on a reinforced beam and slab bridge at Dornie. The Transport Research Laboratory, Crowthorne, 1993, Report RR377.
10. DALY A.F. and RAGGETT S.J. Load test on a jack arch bridge with cast iron girders. The Transport Research Laboratory, Crowthorne, 1991, Report RR310.

©Crown Copyright

Discussion on parts 1 and 2

B. Pritchard, *Consultant to W. S. Atkins Consultants Ltd and Colebrand Ltd*

The first part of my contribution comprises two fanciful options for building new overbridges over live motorways.

The 325 m long Maupré Valley viaduct near Charolles in France was designed and built by Campenon Bernard BTP in the mid-1980s as part of the highways innovative policy of the French Administration for Roads.

The highly original construction is of structural steel and prestressed concrete. The triangular box, 3 m deep, consists of a concrete-filled 610 mm dia. steel tube bottom flange, twin inclined webs of trapezoidally corrugated 8 mm steel plate shear-connected to a closing transversely prestressed in-situ concrete deck slab cast on participating permanent perforated steel formwork. The whole cross-section is longitudinally prestressed by external cables deflected within the triangular girder.

The structure was erected using push-out techniques, the triangular steel beam section of tube, webs and steel permanent formwork being delivered to site in transportable 12 m sections for assembly into the six consecutive spans, maximum 53.6 m. The steel section was pushed out and the top slab cast and prestressed in close succession, a procedure encouraged by the low weight of under 8 t per metre of the 11 m wide deck.

This highly innovative type of hybrid construction could well be used for construction over live motorways. It would not normally be built in push-out fashion for reasons of access, economy and safety, but lifted into place. Fig. 1 shows the proposal, which uses smaller twin versions of the Maupré cross-section erected side by side, with a slab stitch joint formed with either reinforcement or coupled tendons. Initial sizings indicate a total lift of under 170 t for each of the two half-decks, using high strength deck slab concrete.

The smaller girder depth is high enough to allow access to the prestressing cable and for interior painting, inspection and maintenance. An exterior painting gantry is also indicated, adding 200 mm to the effective construction depth. It is proposed that the gantry should be stored in an off-site maintenance compound and hoisted into position from the verges. Alternatively, the construction could be enclosed in GRP or aluminium, although this could add extra expense and detract from the novel and attractive appearance.

Prestressed concrete could be used for another form of erection, the 'drawbridge' method, which also comes from Continental Europe. The idea is based on the Argentobel arch bridge built in Germany, and a French proposal for multi-span viaducts.

DISCUSSION

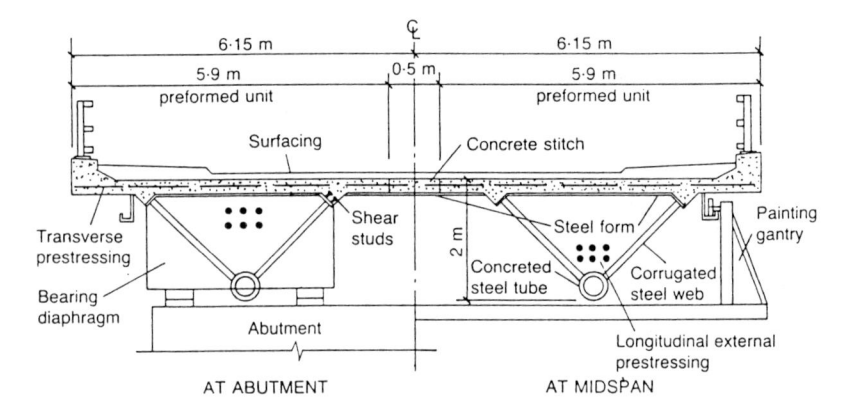

Fig. 1. Composite steel beam/prestressed in-situ concrete deck 45 m replacement bridge

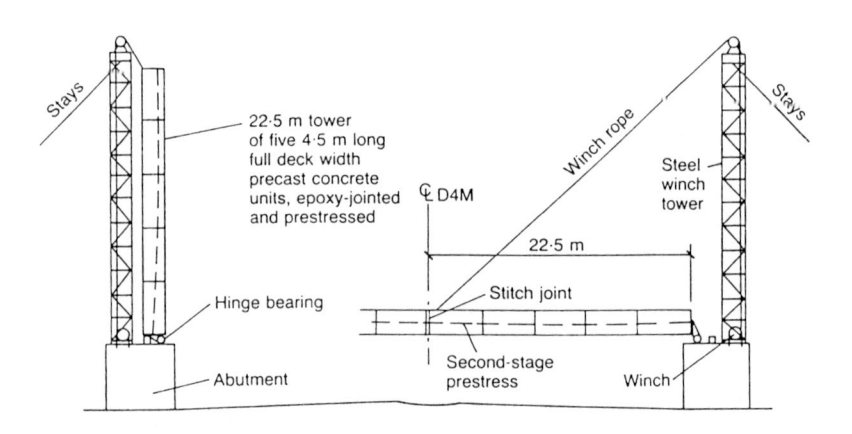

Fig. 2. Prestressed segmental precast concrete 45 m replacement bridge

DISCUSSION

In the original proposal an in-situ concrete box girder is built vertically over temporary hinges on one abutment using slipforming techniques and external prestressing. The tower crane used for concreting is stayed for the vertical rotation into place of the completed deck.

This method, originally proposed for multi-span construction, would prove viable for the construction of a number of similar replacement bridges only where the large and expensive erection equipment could have an economic number of uses. If this were possible, the use of double-leaf drawbridge construction could be considered, with only half-span construction height required over each abutment. A small stitching gap at midspan could be coupled or reinforced and concreted after rotation, with further composite concrete and external prestressing added to cater for surfacing and live loading.

With the reduced height of the drawbridge leaves, it could also be possible to substitute full-width prestressed precast concrete segmental constructions. These various proposals are shown in Fig. 2.

The minimum approximate 37 m and 44 m spans of the main beams for single-span D3M or D4M replacements tend to favour steel construction. Steel beams can be brought to site in braced pairs in readily transportable sections up to the 27.5 m long and 4.5 m wide limits. On site they can be friction-grip-bolted to full span length prior to lifting into place. A typical solution for a two-lane side road crossing is the single 45 m span devised by Alan Hayward, with a commendable construction depth of only 1.89 m. The full-width deck slabs are precast in short lengths and placed over the three previously erected steel-beam assemblies, the pockets in the slabs fitting over clusters of shear connection welded studs. Stitch joints between slabs are merely filled with in-situ concrete, the stitch being unreinforced.

It is suggested that the procedure could be improved further by fixing the shear studs through the pockets after slab placement. This would permit a push-out assembly of the slab units from one or both ends of the bridge, with no disruptive lifting into place by cranes from the widened road.

Another improvement would lie in the provision for future maintenance painting, with minimum, if any, traffic disruption. This could be accommodated by using GRP or aluminium enclosures or by providing a painting gantry, adding 200–300 mm to the required construction depth.

A possible weakness of the deck slab joint is that differential temperature and wheel load effects could cause joint cracking. If the waterproofing is good enough, the cracking could be accepted. A more positive approach would be to provide wider precast slab units with lapped or coupler-connected longitudinal reinforcement. Fig. 3 shows a lapped joint alternative, using in-situ topping composite with a dished length of precast slab in the lap locations. The push-out roller bearing and levelling jacks assembly is also shown

The span/depth ratios quoted in the excellent M1 and M6 papers could be improved further by adopting an innovative method from Belgium which adds prestressed composite beam prestressing to the pre-bent Preflex bridge beam used in the UK in the 1960s and 1970s. The new beam is shown in Fig. 4 and consists of a pre-bent steel beam encased in hybrid prestressed concrete.

The highly complex assembly consists of the following stages. First, the steel beams are placed on a pre-tensioning bed and are prestressed by wires (a) attached to the bottom flange. Second, the prestressed steel beam is preflexed (i.e. bent

DISCUSSION

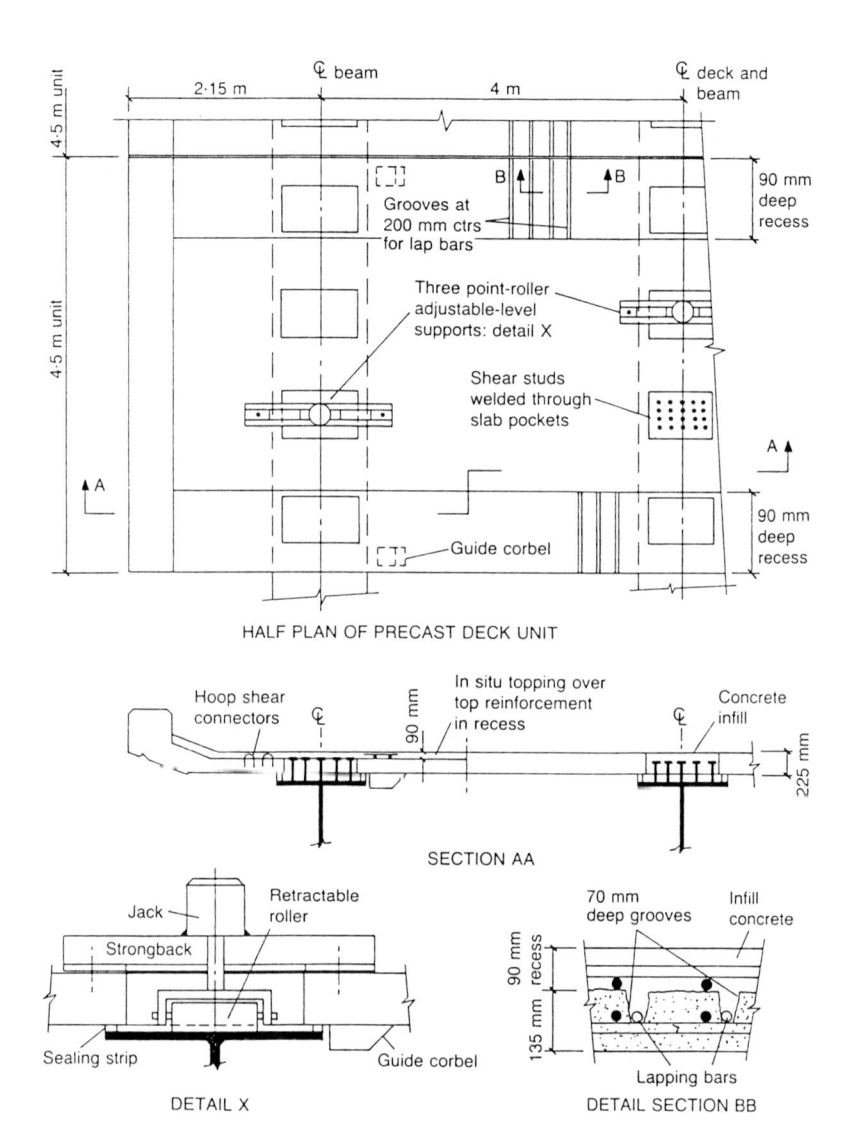

Fig. 3. Assembly and stitch jointing for precast concrete deck slabs

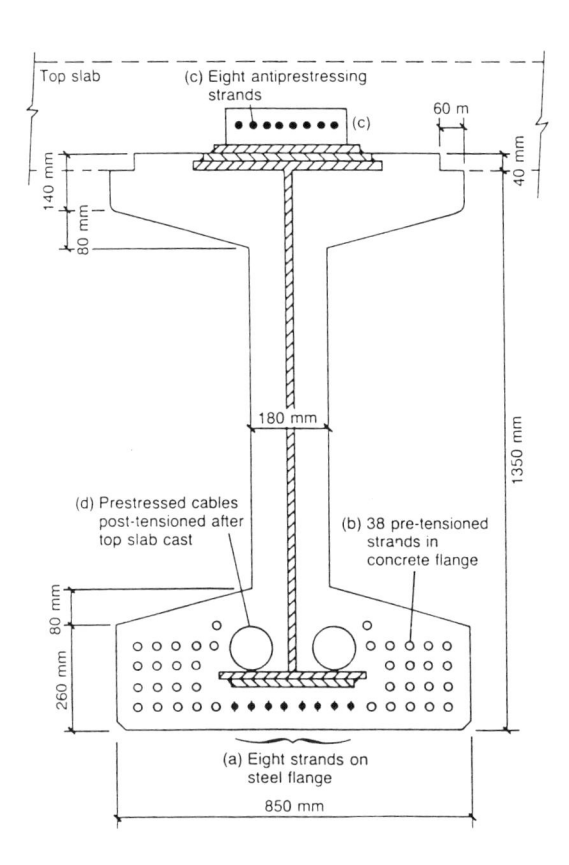

Fig. 4. Prestressed and preflexed steel–concrete composite beam: span 47 m, weight 90 t; overall depth 1.6 m

DISCUSSION

downwards) by external jacks. Next the pre-tensioning strands (b) are stressed by the pre-tensioning bed jacks. Then the beam and associated prestressing strands and cables are concreted and cured. Next the antiprestressing strands (c) attached to the top flange and outside the concrete are prestressed. Then the preflex loading of the second stage is removed by the release of the external jacks. Following this, the pre-tensioning is released on to the composite beam. Next the beams are transported to site and erected; the formwork for the top slab reinforced concrete is hung from the beams. Then the antiprestressing strands (c) are released and removed. Next, the top reinforced concrete slab is cast and finally the post-tensioning cables (d) are prestressed.

The beams were used for a viaduct crossing the River Meuse near Liège in Belgium, with spans of 47 m. A remarkably low construction depth of 1.55 mm, giving a span/depth ratio of 31, demonstrated the high strength of these complex beams and their possible suitability for replacement bridges in the UK.

Unless special transport arrangements were possible, it would be necessary to bring the steel beams in sections for assembly by site welding. All the deflection and prestressing beds and equipment would also have to be site-based. No doubt this could be accepted where a large number of replacement bridges would justify the setting-up of a central site adjacent to the motorway for beam production.

T. J. Upstone, *Manager Engineering, Redpath Dorman Long (Retired)*

The current motorway widening programme calls for approximately 1500 bridges to be replaced or severely modified. Most of these bridges are over the motorway, and the only practical solution is their complete replacement. So far, such operations have been carried out using specialist plant such as cranes of 600 t capacity and trains of multiple 250 capacity bogies carrying complete bridges along the motorway. The total reliance on this unique equipment has very serious consequences if it breaks down during operation. The use of blasting for demolition of the old bridges is not entirely predictable. But even when all goes well, all of these methods involve complete stoppage of the motorway traffic on numerous occasions at each site. Further, the removal of the old bridge, its piers and abutments, and the building of the new piers, if required, and the erection of the new bridge, all in close proximity to live motorway traffic adds considerably to the time required, the costs and the risks involved.

An alternative method of replacing motorway and trunk road overbridges is proposed. An arrangement is shown in Fig. 5 for carrying two lanes of motorway traffic in each direction over the site of an overbridge by means of twin temporary viaducts. This enables motorway traffic to pass above the site unhindered during the demolition and rebuilding of the bridge. They are built in sequence from access roads outside each hard shoulder that would later be incorporated into the widened carriageways. The viaducts occupy the hard shoulder and slow lane of each carriageway. Temporary roads crossing the now 'dead' motorway beneath the viaducts join up the slip roads either side, allowing continuous traffic on the 'minor' road.

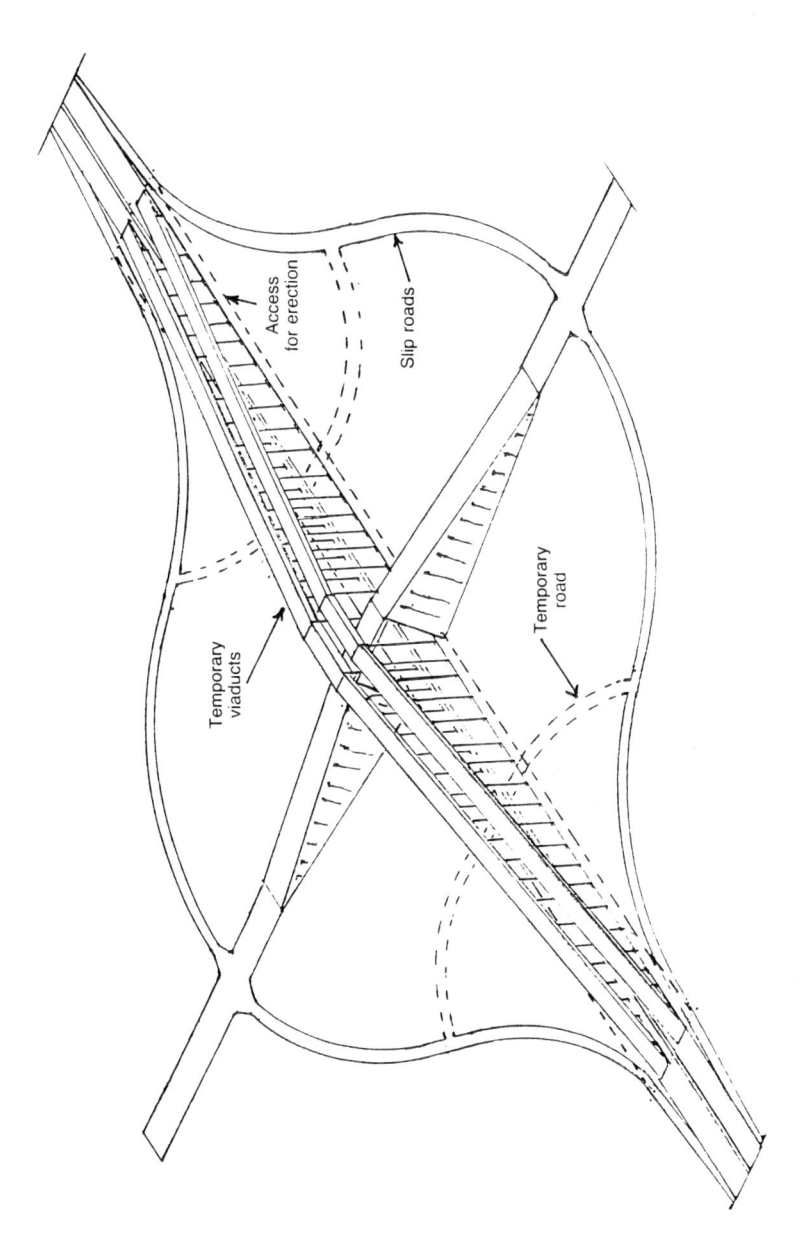

Fig. 5. Replacement of dual two-lane bridge over dual three-lane motorway using dual two-lane temporary structures and temporary diversion roads and existing slip roads

DISCUSSION

The area of motorway beneath the viaducts and adjacent to the overbridge becomes a working area for the contractors to demolish the old bridge and rebuild the new one unhindered, and by the most economical method.

The same methods can be used when the motorway is in a cutting, as in Fig. 6. In this case it may be necessary to provide a temporary bridge over the motorway and alongside the overbridge for the minor road traffic. Such a bridge would pass under the twin temporary viaducts and would be built after they had started to carry the motorway traffic.

A cross-section of the motorway (Fig. 7) shows the temporary trestle viaducts before and after widening. The trestles are carried on concrete spread footings placed directly on the existing road surface. Suitable bridge spans, prefabricated in lane width and various span lengths, together with supporting trestles are available in the UK for hire or purchase. The viaducts would be used many times on a given length of motorway and then dismantled and taken to other contracts to be used again and again.

An actual motorway junction is shown in Fig. 8 where there would be two overbridges to be replaced. The motorway is in a cutting and even the roundabout road is below the general ground level. The twin temporary viaducts are carried over both the bridges. The length of these viaducts depends on the motorway longitudinal profile and the slope chosen for the viaducts. The longitudinal profile is shown in Fig. 9, and with a slope of 4% the viaducts are 800 m long. Increasing the slope to 5% would reduce the length to 710 m. For a junction such as that shown in Fig. 8, the two overbridges would be replaced in sequence. The roundabout carriageway would be made two-way temporarily so that one bridge could carry all the cross traffic while the other was rebuilt, and then vice versa. The two-way traffic round the roundabout could be controlled by synchronized traffic lights or, alternatively, it may be possible to construct temporary mini-roundabouts at each secondary junction as shown in Fig. 10.

Among the advantages of this system are the following.

(a) Continuous two-lane motorway traffic is maintained in both directions during the building and dismantling of the temporary viaducts and at all stages throughout the work on the overbridges.

(b) The contractor has full possession of the motorway carriageways in the vicinity of the overbridges and a large working area adjacent, all with good access from the motorway slip roads or the crossing roads.

(c) The old bridge, its abutments and piers, can be demolished and the new bridge abutments and piers can be carried by the most economical, traditional means with no adjacent live motorway traffic to contend with.

(d) The new bridge superstructure, in steel or concrete, can be built in situ using falsework below if necessary. Small conventional cranes and other equipment can be used and can be readily replaced in the case of breakdown.

(e) Using conventional methods of demolition and rebuilding the bridge means less possibility of hold-ups and is more predictable.

(f) Suitable bridge spans and trestles for the temporary viaducts are available and would have multiple uses on the many stretches of motorway to be widened.

(g) The same methods could have equal advantages on symmetrical, asymmetrical and parallel widening schemes, all of which have at least four lanes or hard shoulders in the same positions for both the existing and widened motorway.

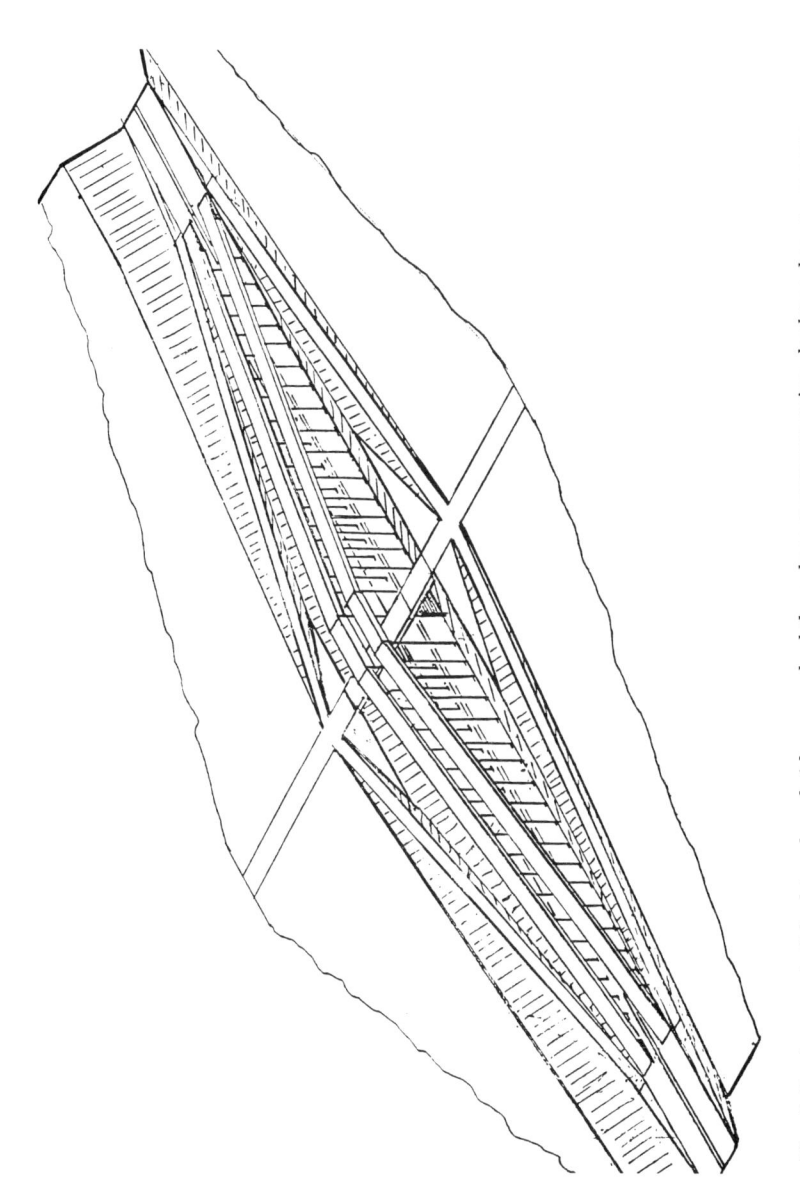

Fig. 6. Replacement of dual two-lane bridge over dual three-lane motorway using dual two-lane temporary structures

DISCUSSION

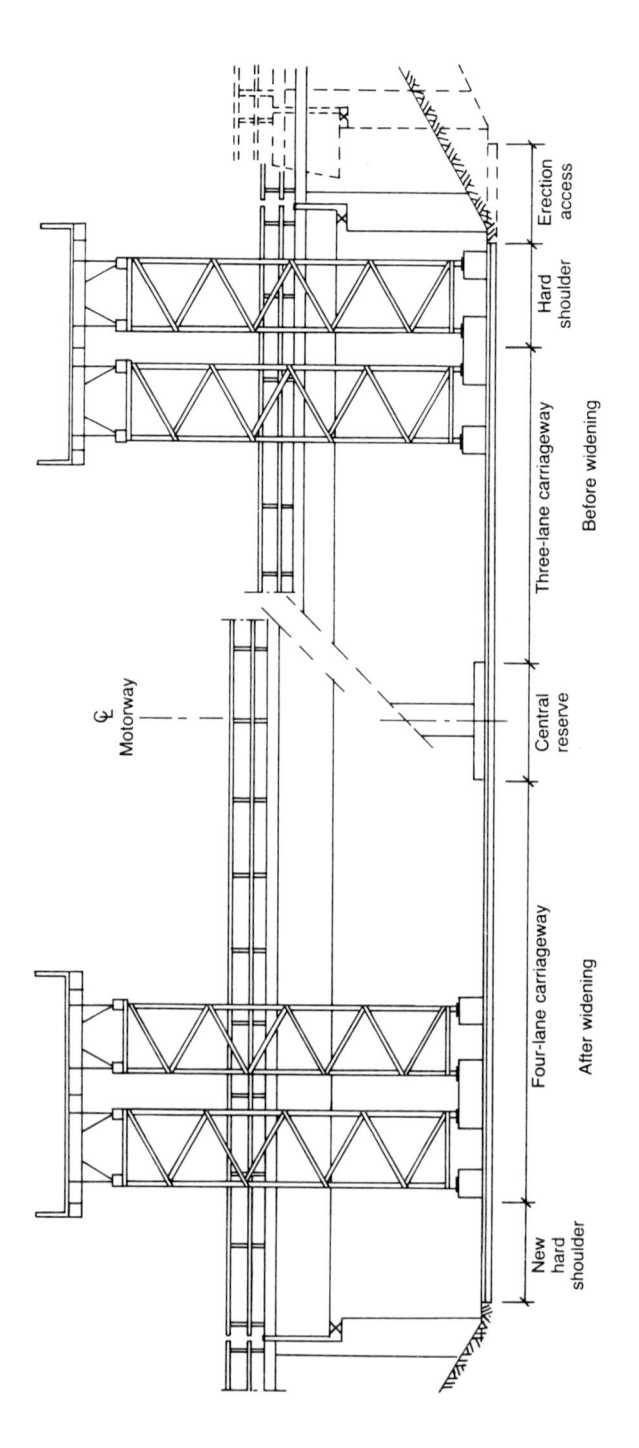

Fig. 7. Replacement of motorway overbridges

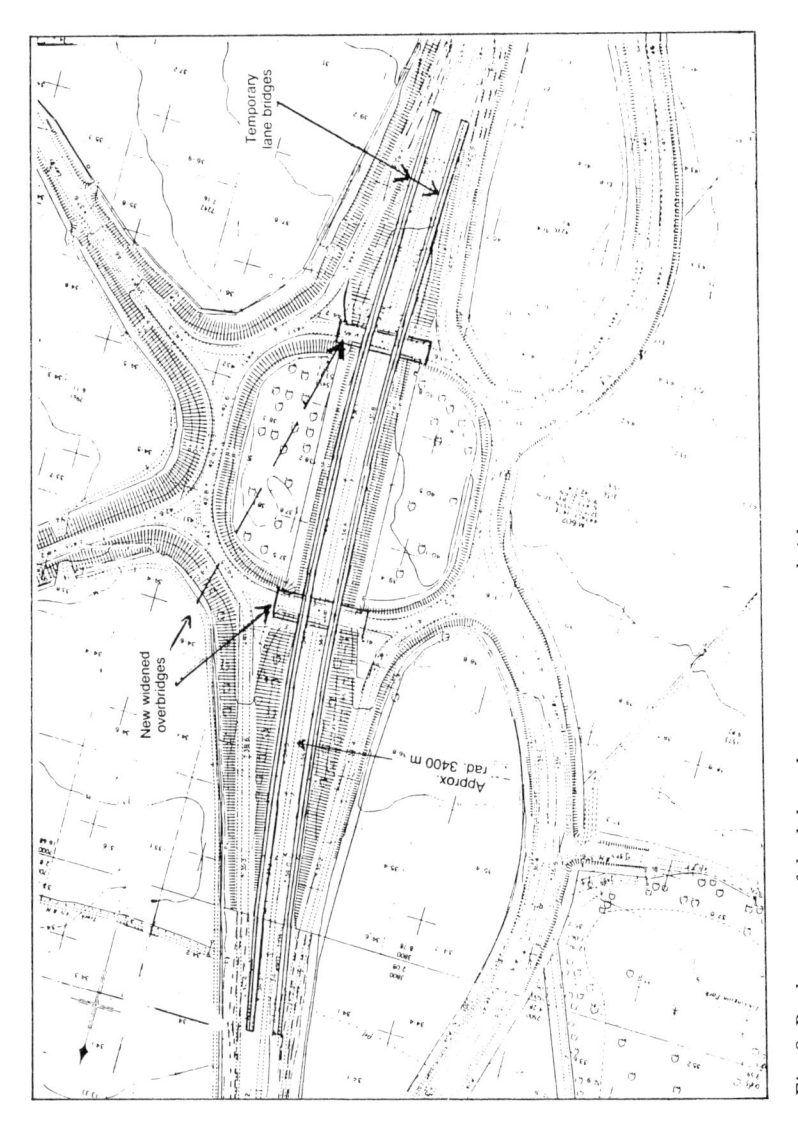

Fig. 8. Replacement of dual three-lane motorway overbridges

DISCUSSION

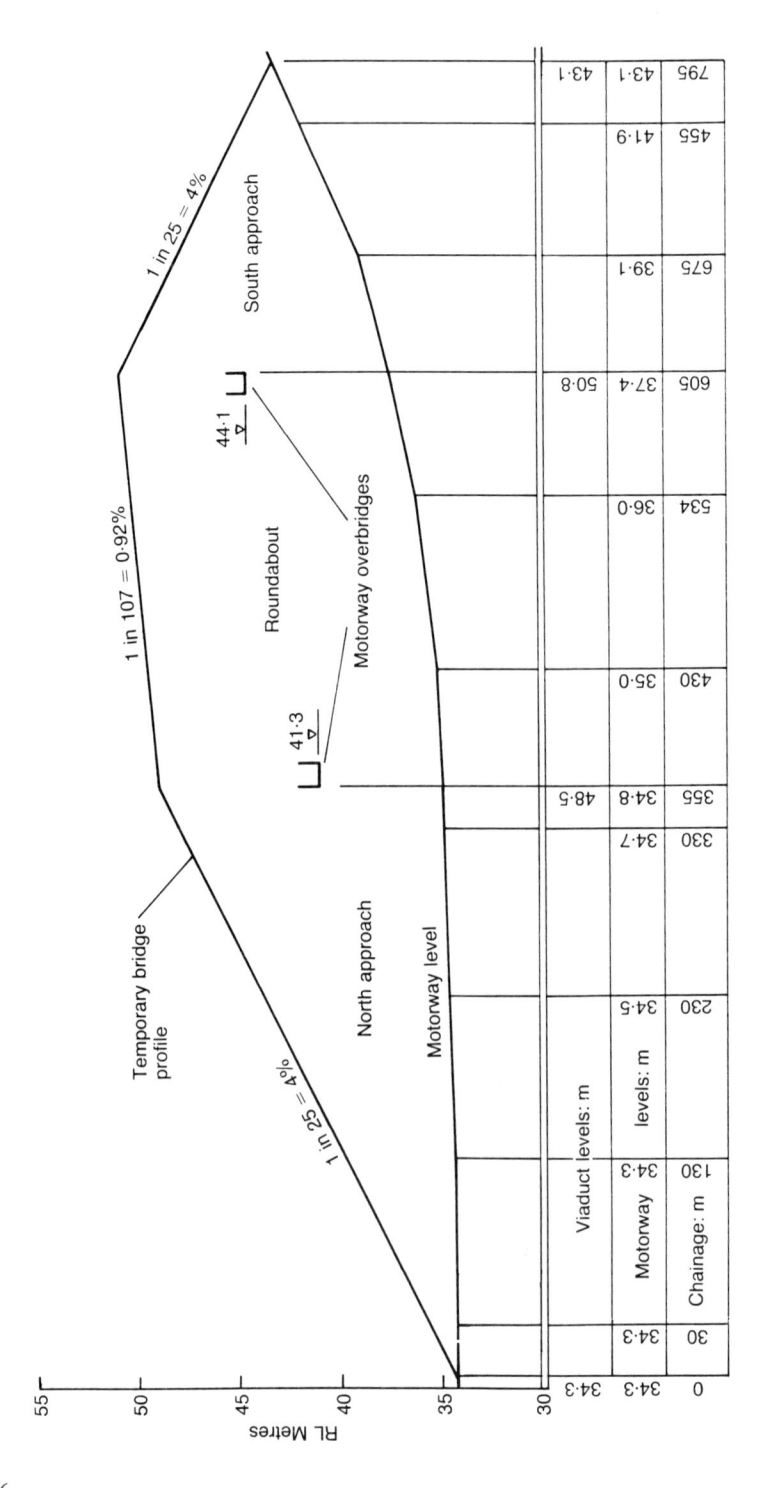

Fig. 9. Replacement of motorway overbridges: developed section on east carriageway

DISCUSSION

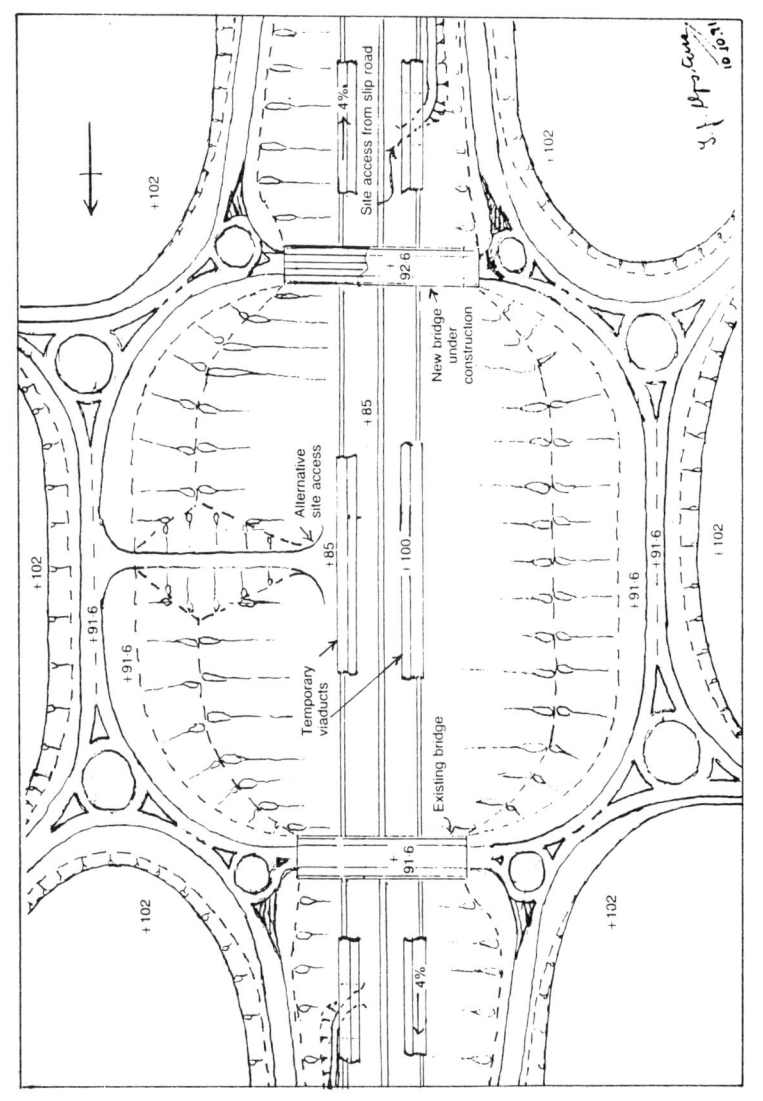

Fig. 10. Traffic arrangements replacement of south bridge

DISCUSSION

The foregoing describes a safe method of replacing overbridges without interruption of the traffic on the motorway or adjacent roads. It is believed to compare favourably both in time and cost with the methods that have been used up to the present time.

W. C. Arrol, *Department of Transport*

We hear today about all the problems involved in the alteration of bridges. When these bridges were built they were generally tailor-made for the particular road arrangement that was required. Spans were the minimum necessary; cross-sections had nothing in hand. In other words, no thought was given to the future.

Some types of structure are suitable for alteration and some are not. In retrospect we should have constructed far more of the former type. A type of bridge extensively built, which is practically impossible to widen, is the prestressed concrete spine beam bridge with large elegant cantilevers. Had these been built with wider beams, smaller cantilevers and some reserves of strength at critical sections, new extended cantilevers could now be constructed to provide additional width. This requires major work, although far less than if the whole bridge were being reconstructed. This type of thinking can be extended into many other types of bridge form.

The point I wish to make is that with the reconstruction programme now being undertaken in many areas, we should learn from the past and not repeat what has gone before. New designs should have some features built into them which will enable our descendants to modify the structures rather than to demolish them. Designers will not be able to anticipate all the changes that might occur, but if they can cover some of them a better situation will exist. I think this will provide better engineered bridges, for a minimal increase in cost, that can expect to have longer lives than those described in the papers.

I should like also to make an observation about one of the bridges in Mr Champion's Paper.

The choice of relocating Green Street Green Bridge rather than replacing it was made mainly because a suitable method of demolition, that was acceptable to the Department of Transport (DoT), could not be found.

The A282 is part of the M25 ring and traffic flows are very high. At Green Street Green Bridge there was no suitable diversion route for the traffic and therefore closures, other than short ones overnight, were unacceptable. Conventional demolition could not be carried out in the available time and there was little experience of removing a post-tensioned structure from extensive temporary works across the carriageway above deck level. Relocating the structure was seen as the only feasible acceptable solution.

A specialist subcontractor was brought into the design process and this resulted in a well thought-out scheme which was successfully completed on site. It was an expensive operation, dictated by site conditions, but the cost was comparable to demolition, had this been possible, and to constructing a new bridge.

DISCUSSION

P. F. Johnson, *Maunsell Group*

It is said that steel beam bridges are very lively. Also, buried joints are very popular. Could Mr Dawe say whether or not the testing programme on joints now being carried out at TRL includes the testing of the rotational response time of buried joints. Will they react to the deflections of passing HGVs?

D. Asprey, *Avon County Council (Highways and Transportation)*

Mr Dawe has referred to the high risk that bridges will be hit at some time during their life. I suggest that this is of particular concern in the case of relatively light bridges, and that for these, the DoT should consider providing fail-safe measures to prevent a catastrophic situation from occurring. History teaches that if an accident can recur it will do so sooner or later; engineers are obliged to safeguard against this possibility.

S. J. Matthews, *Associate Director, Bridges Division, Frank Graham Consulting Engineers*

I was a little concerned about the results of the fatigue tests on brick masonry given in the Paper by Mr Broomhead and Mr Clark. They appear to suggest that a limited number of high stress loading events (such as may occur under an increased lorry axle load) could lead to a rapid deterioration of a structure which had previously appeared satisfactory. Could Mr Clark comment on this, and could he confirm whether or not the test pieces were constructed from mortars and bricks representative of those currently existing in masonry arches?

Dr A. Mahdi, *Essex County Council*

In one of the jack arch bridges, Dr Cullingham shows a non-linear load-deformation using grillage analysis. How did he decide on the level of the stiffness used?

Mr Dawe

In reply to *Mr Johnson*, the DoT is developing two types of joint testing machine: one will enable a wheel load to be run repeatedly over the surface of a sample joint; the other will allow repeated translational and rotational displacements to be applied to the edges of a joint which can also be held at specified temperatures. The intention is to develop suitable performance criteria for different types of joint

DISCUSSION

which are representative of the movements and loads which a joint might actually experience in service. Any joint intended for use on the DoT's roads would then have to demonstrate that it could satisfactorily endure an appropriate number of specified load and/or displacement cycles when tested in the machines.

In reply to *Mr Asprey*, one item of the 15 year programme covers the assessment procedures and remedial measures which are required to ensure that existing structures have adequate resistance to withstand the impact of errant or overheight vehicles. However, by their very nature it is unlikely that lightweight structures will have sufficient impact resistance built into their own construction. In such cases it is possible to reduce the possibility of collapse by alternative measures such as by protecting supports with higher containment safety barriers or by providing increased headroom. However, it must be recognized that there will be many existing lightweight structures which cannot be protected completely by such means and which will have a higher risk of failure after a collision than heavier structures.

Mr Boyes

In the past, motorways and other roads were often built on the assumption that they were a once and for all solution to the problem of inadequate traffic capacity and congestion. Experience has shown this to be an inflexible stance and, as a consequence, we are busily engaged on widening our present motorway system. Creating the right level of flexibility is one very important reason why we need to consider the future when designing our bridges.

The demolition of bridges, most of which are less than thirty years old, in the present programme of widening must be seen as an inefficient use of resources. It can also be enormously disruptive to the traffic flow, which has to be maintained on a live motorway. We can avoid this by building in flexibility and I am therefore fully in agreement with *Mr Arrol*'s views on giving thought to future needs. As I indicated in my Paper, this does have to be balanced by an economic evaluation to show that sensible choices are being made.

We tend to use the term 'widening' in two ways when addressing bridge modification. We can use it to mean extending the length of the bridge span where it crosses over an increased carriageway width; and we can use it to mean broadening the width of a bridge in order for it to carry the increased width of carriageway, which is perhaps a better use of the term. Mr Arrol has drawn attention to this latter aspect and, in the light of his remarks, it is apparent that we shall need to cover this more fully at the next Bridge Modification Conference. Nevertheless, I consider that the principles enunciated at this Conference apply equally and that, used sensibly, they would lead to soundly based proposals.

DISCUSSION

Mr Champion

Mr Arrol's remarks provide valuable additional information to the decisions behind what remains one of the most unusual bridge modification exercises yet completed.

Mr Clark

My reply to *Mr Matthews* is as follows. The brick panel tests showed that brick masonry can deteriorate quite rapidly under loads that are less than the ultimate compressive stress, especially if the bricks are wet. It must be emphasized that the capacity of a brick arch may not be limited solely by masonry compressive stress; in most cases, the failure mode is as a mechanism, and the masonry strength in such cases is less important. However, if an arch is built from fairly weak masonry and the barrel is saturated, then the structure could well be at risk from premature failure as a result of heavy lorry axles. It would be prudent in such cases to carry out an analysis of the arch to check the stresses under the heavy axles and to test the masonry strength.

With regard to the bricks and mortar used in the tests, the bricks were of a type manufactured for use in repairing old structures. They have similar properties to old bricks and are weaker than modern bricks. The mortar used was a cement-lime mortar that was designed to be of similar strength to old mortars. The reason for using new bricks was to ensure that samples had consistent properties and this was verified by crushing several piers to test this consistency.

Dr Cullington

In reply to *Dr Mahdi*, non-linear load-deformation curves were obtained directly by experiment, rather than by grillage analysis, on three bridges. The three bridges behaved in a similar fashion, all providing evidence of composite action between the girders and the fill. Non-linear behaviour was attributed to the reduction of composite action and the yielding of the girders as the load was increased and the ultimate limit state approached. For one of the bridges, a linear grillage analysis was carried out to quantify the transverse distribution of load (expressed as proportion factors) in the linear range and to allow comparison with the distribution at the ultimate limit state as determined by experiment. Initially, the properties of the grillage model were obtained by calibration, using the response of the bridge obtained in vehicle load test.

Investigation of distribution in the non-linear range was possible because the bridge consisted of two identical spans which could be tested to collapse separately. The first span was loaded using a rigid knife edge acting across the full width of the bridge. All the beams were thus constrained to act uniformly so that an average could be found.

The second span was loaded using a simulated bogie which caused a non-uniform transverse distribution. In the linear elastic range, proportion factors

DISCUSSION

were derived directly from the measured soffit strains. In the non-linear range, the load in each beam was found by substitution, using the average beam properties obtained from the knife edge test.

The outcome was that the grillage model found by calibration from the vehicle test gave proportion factors that were acceptable up to the ultimate limit state, although the absolute values of displacement and strain were invalid as would be expected. There was some favourable redistribution as the ULS approached but this was not significant. More importantly, there was no unfavourable redistribution which might have occurred had the transverse properties of the jack arches deteriorated at high loads.

It was not the intention to extend the grillage analysis into the non-linear range. To do so would require the selection of member properties outside the elastic range, with the attendant difficulties implied by Dr Mahdi's question.